# The Afterlives
## *of* Animals

*A Museum Menagerie*

EDITED BY SAMUEL J. M. M. ALBERTI

*University of Virginia Press* | *Charlottesville and London*

University of Virginia Press

© 2011 by the Rector and Visitors of the University of Virginia

"The Afterlife of Chi-Chi" © Henry Nicholls

Printed in the United States of America on acid-free paper

First published 2011

9 8 7 6 5 4 3 2 1

Library of Congress Cataloging-in-Publication Data
The afterlives of animals : a museum menagerie /
edited by Samuel J. M. M. Alberti.
p.   cm.
Includes bibliographical references and index.
ISBN 978-0-8139-3167-8 (cloth : alk. paper)
ISBN 978-0-8139-3208-8 (ebook)
   1. Famous animals—Biography.   2. Famous animals—
Collection and preservation.   3. Animals—Biography.
4. Zoological specimens—Collection and preservation.
5. Taxidermy—Anecdotes.   6. Animals and civilization—
History.   7. Menageries.   I. Alberti, Samuel J. M. M.
QL793.A38   2011
590—dc22                                    2011008110

# Contents

# Acknowledgments

Many people and institutions helped to bring this volume to fruition, and specific thanks can be found in the pages that follow. A number of organizations were generous with resources, staff time, and images, including Bristol City Museum and Art Gallery; the British Museum; Chetham's Library, Manchester; the Design Museum; Glasgow Museums; the Great North Museum: Hancock; the Hunterian Museum and Art Gallery, University of Glasgow; the Manchester Museum, University of Manchester; Marine Connection (www.marineconnection.org); National Museums Scotland; the Natural History Museum, London; the Wellcome Library, London; and the Yale Center for British Art.

As editor, I could not have asked for a more dedicated, enthusiastic, and amazingly punctual group of contributors. Former colleagues and students at the Centre for Museology suffered my zoological enthusiasms with forbearance. Elsewhere, I have been especially aware of our debt to colleagues and friends including Ebony Andrews, Steve Baker, Fay Bound Alberti, Anna Bunney, Malcolm Chapman, David Craven, Roy Garner, Julie Harvey, Antonio Marcelo Herrera, Angie Hogan, Rebecca Machin, Henry McGhie, Susan McHugh, Steve McLean, Nick Merriman, Irit Narkiss, Adrian Norris, Michael Powell, Maggie Reilly, Liz Sandeman, Hanna Rose Shell, Mark Steadman, Phyllis Stoddart, Liv Emma Thorsen, Myna Trustrum, Clare Valentine, Mick Worboys, and Boyd Zenner. Some of these colleagues were present when the contributors met at the Manchester Museum in December 2009 to present and discuss their contributions. We are grateful to all those who participated in the workshop, which was generously supported by the Manchester Museum; the University of Manchester Centre for the History of Science, Technology and Medicine; and Renaissance North West.

SAMUEL J. M. M. ALBERTI

# Introduction

## The Dead Ark

One of my favorite natural history exhibits, now sadly extinct, was "Abel's Ark" at the Hancock Museum in Newcastle upon Tyne (now part of the Great North Museum). Faced with a collection of sporting trophy heads inherited from a local Victorian naturalist, twentieth-century curators built an educational display in which the decapitated mounts poked their heads through the window of a jolly ark (see fig. 1).[1] Few displays demonstrate so effectively in a single glance the changing functions of natural history museums and the radical shifts in the meanings of animals: from life on the savannah to a sportsman's prize, from hunting mount to specimen to educational object.

Elsewhere in the serried ranks of other natural history museums, there are many animals that similarly refuse to be constrained by their zoological classification. Their fame in life and their iconic status in death defy taxonomy. They are not only specimens, but also personalities; not only data, but also historical documents. *The Afterlives of Animals* selects some especially interesting examples and traces their individual histories both before and after death: their movement and meanings in life, how they died, and then how they came to be in collections. By assembling a series of such "animal biographies," we are able to trace the shifting meanings (scientific, cultural, emotional) of singular animals and their remains. The stories that follow take us from fields and rivers through zoos and menageries to museums. From Balto the husky to Chi-

FIG. 1. "Abel's Ark" at the Hancock Museum, 2005. (Great North Museum: Hancock and the Natural History Society of Northumbria/Steve McLean)

Chi the panda, they show us how people relate to animals in life and death. For many people in the modern West, these zoological shows and collections were the principal (or indeed exclusive) site for their encounter with the material animal, and the herd of beasts in these pages includes representatives of some of the most popular exhibits. As Nigel Rothfels has observed: "The perennial stories of whales, elephants, pandas, and other charismatic species, make clear that the stakes in representing animals can be very high. Who controls the representation and to what ends it will be used [is] of profound importance."[2] It is the premise of this volume that those who have custody of dead as well as living animals play crucial roles in this representation, but that their intended interpretations do not always tally with the meanings afforded to animals by visitors.

It might seem strange to write biographies not of people but of animals, and even stranger to extend these biographies beyond death, to the preserved remains of these animals that might simply be considered "things." But in approaching the singular histories of specific animals, whether as individuals or as objects, we draw on techniques deployed for some time by anthropologists and others who are interested in the "social lives of things," as Michelle Henning and other contributors indicate in their essays below.[3] Like these scholars of material culture, we approach the following studies through the trajectories of specific items and the relationships they form with people and other objects. As one would when narrating a human biography, we account for key moments in our thing's life and afterlife. Where did it go, whom did it meet? How has its status changed? What makes it different from other, similar things or individuals? Because of the nature of our subjects, however, we acknowledge that there are periods and spaces in their trajectory that cannot be recovered, that these "biogeograhies" (see the essay by Merle Patchett, Kate Foster, and Hayden Lorimer) are often messy and unchartable. Furthermore, in taking animals as our protagonists, we follow a venerable tradition, that of the "it-narrative," an eccentric subgenre evident from the eighteenth century in which the story was told from the perspective of objects or animals.[4]

Here we are especially interested in the fate of these animals after death: their *afterlives*. Just as animals were mobile, flexible entities with changing meanings as they traveled through sites for living display, so this accrual continues postmortem. We mean "afterlife" not in the spiritual way, clearly, but rather in the same sense as the conservator Richard Jaeschke in his discussion of the fate of archaeological objects: "For the archaeologist, the life of the object is fixed at the moment of its discovery [a notion that could equally be applied to the biological death of an animal for the museum zoologist]. For conservators, however, the situation is not so simple and the life of the object continues." The last of the stages in the career of objects that Jaeschke proposes, somewhat tongue-in-cheek, is the afterlife: during which "the object is placed in its heavenly home, the museum or archive, to remain for eternity . . . [but] the afterlife seldom proves a heavenly resting place and usually involves the object in more adventures and perils. It does not cease to age and to have a history, but its history of use has become a history of treatment."[5] As Geoffrey Swinney reflects in his afterword to this volume, the animals herein have many different kinds of afterlives, in different places and in different media. Afterlives, Swinney shows, are created in the recounting—we ourselves are contributing to this process in this book.

Written by contributors from different disciplines and backgrounds about different species of animals at different times, the accounts that follow nonetheless exhibit some striking parallels. In what remains of this introductory piece, I draw attention to two broad themes in particular: the journeys they made, whether geographical or physical; and the accompanying shifts in meaning. The biblical title of this introduction, as well as drawing attention to the Hancock Museum's imaginative reuse of mounts, is intended to encapsulate both of these themes: transition (Noah—arguably the first collector—saving and transporting animals), and new meaning (such as the metaphorical use of the ark by modern zoos in conservation mode).[6]

## TRANSITIONS

The list of places visited or inhabited by the animals here would be as long as the book itself, including several countries in at least four continents; rivers, fields and Arctic wilderness; a royal menagerie and a royal college; several zoos, workshops, storage facilities, and numerous museums; as well as all the places we have been unable to recover. And thanks to our biographical perspective, tracing the paths of animals in life, preparation, and exhibition (or not), we are able to show the dynamic links between these different spaces for the experience and display of animals. Furthermore, as shown by most essays herein, representations of particular animals then traveled even farther in print, image, film, and online.[7]

Whether in body or not, in moving between these sites, these animals circulated understandings of the natural world. As I argue in my essay on Maharajah, massive beasts were in effect chunks of landscape ripped asunder and transplanted to urban locales.[8] They and their smaller peers were the embodiment of what the historian of science James Secord dubs "knowledge in transit." Secord advocates breaking down the purported boundaries between the practice of science (how scientists understand nature) and the popular, "public understanding" of science (how the rest of us understand nature). Instead we should think more holistically about the circulation of knowledge, about the sciences as a series of "communicative actions."[9] Scientific knowledge is not disseminated simply by virtue of its truth value, but rather spreads unevenly and contingently. This circulation, I would argue, involves not only concepts but is also (and especially effectively) channeled through things. Are not the animals we write about in this volume *material* knowledge in transit, bringing experiences of nature with them to different sites and audiences?

One transition of particular interest in the essays that follow is between zoos and museums.[10] Although specimens that were extracted from their so-called "natural" habitat may well be valued more as markers of nature, many of the zoological specimens in natural history collections arrived from the local zoological gardens, as we see in the following biographies of Chi-Chi the panda, Alfred the gorilla, the elephants Maharajah and Sir Roger, and the unnamed mandrill. Acquisition routes then extended farther, to traveling menageries, to the peripatetic whale shows that Henning mentions, and even to circuses. These connections, like those to hunting upon which Garry Marvin reflects, are perhaps associations that museums would rather not emphasize, seeing them as somehow "tainting" the scientific data. But the link to zoos is interesting for our purposes because it allows for the unusual possibility—pet taxidermy notwithstanding—that the same audiences might have experienced particular individual animals in both life and death, and allows us to compare and contrast their meanings.

Even more central to *The Afterlives of Animals* than the connections between zoos and museums is the transition between life and death. The writer and naturalist Henry David Thoreau grumbled:

I hate museums; there is nothing so weighs upon my spirits. They are the catacombs of nature. One green bud of spring, one willow catkins, one faint trill from a migrating sparrow would set the world on its legs again. The life that is in a single green weed is of more worth than all this death. They are dead nature collected by dead men. I know not whether I muse most at the bodies stuffed with cotton and sawdust or those stuffed with bowels and fleshy fibre outside the cases.

Where is the proper herbarium, the true cabinet of shells, and museum of skeletons, but in the meadow where the flower bloomed, by the seaside where the tide cast up the fish, and on the hills and in the valleys where the beast laid down its life and the skeleton of the traveller reposes on the grass? What right have mortals to parade these things on their legs again, with their wires, and, when heaven has decreed that they shall return to dust again, to return them to sawdust? Would you have a dried specimen of a world, or a pickled one?[11]

This, then, is the tension between museums and zoos, and within museum themselves. In investigated and collecting living nature, museums become mausolea, storehouses of millions upon millions of dead things in drawers and jars, grim reminders of mortality. They comprise, as Benedict Anderson put it, a necrological census.[12] This is why taxidermy is so interesting, as we see in the essays that follow on Sir Roger, Balto, Alfred, Chi-Chi, and hunting trophies. Whereas skeletal mounts, wet specimens, and study skins such as the mandrill and the hen harrier are clearly dead, taxidermy is intended to give the illusion of life.[13]

And so—in this context at least—death is not the event horizon we might assume it to be. In taxidermy as well as other preservative media, as Richard Sabin shows so vividly in his essay, the biological death of the living beast is the birth of the specimen. The causes of death mentioned in this volume range from "natural" (such as tuberculosis) to violent (especially gunshot). But for those beasts destined to become museum specimens, biological death is only one moment, one narrative hinge of many (admittedly a particularly resonant one) in the life/afterlife of the animal. Contrary to the philosopher who claimed of one object that was once alive, "This parrot is no more! He has ceased to be! This is an *ex-parrot!*" it *was* still a parrot, but in a different phase of its existence.[14] And they continue to have much in common. Materially, they have many of the same properties, even if only a fraction of the living specimen has remained (whether skin only or bones only). In life they are acted upon, circulated, and displayed, and so, too, in death. Taxidermy mounts continue to stare back at us, albeit from glass eyes.[15]

And yet they are "remnant models," surface without substance, no more complete than the skeletal specimen; and those specimens that look back are in a tiny minority—as discussed in essays below on Chi-Chi, the mandrill, the hen harrier, and Maharajah.[16] As Mark Alvey argues: "It is the animal, and yet it is not. . . . A 'stuffed' panda in a diorama is more real than a photograph, and yet somehow less real than an example of the same species in the zoo."[17] The vast majority of specimens are not prepared for display, however: they are rarely looked at, their aspect is strikingly different from the living beast, and their afterlife is almost a secret (as we see in Sophie Everest's essay).

Whether or not it is lifelike, the end result of the passage from life to death (and the subsequent survival of the remains) involves considerable work —techniques that aid the passage into an afterlife. The death of the animal may or may not have been the result of direct human intervention (see Garry Marvin's essay), but a sustained afterlife certainly is. The first postmortem act was commonly fragmentation, the separation of flesh, bone, and skin—

detailed accounts of which for Chi-Chi, Sir Roger, and especially the Thames Whale can be found below—some parts kept, some discarded, and often different parts going to different places. Thereafter, a range of techniques were deployed to reconfigure and preserve the remains, to render them stable and legible.[18] Conservators, taxidermists, and curators labored on specimens as whalemen labored on their bounteous catch, as Henning shows in her essay: in making a museum object, they "manage the material." Making animal afterlives is an active process, as Merle Patchett, Kate Foster, and Hayden Lorimer show, and upon which Swinney reflects in his afterword. Not only are animals recomposed (so that they do not *decompose*), but they are also embellished, reconfigured in new and interesting ways.

MEANINGS

Capture, transit, display, work, and embellishment layered many different meanings around the animals detailed in these pages. We know from the thriving animal studies literature that animals can be widely polysemic according to context and mode of engagement; like the essays in the Animal Studies Group's *Killing Animals,* in *The Afterlives of Animals* we explore how the transition from life to death impacts further upon these diverse meanings.[19] Specifically, when taken together, these essays demonstrate the variance of understandings and experiences between taxa, between modes of preparation, and, most diffusely, between different "ways of looking."

It will come as no surprise that different kinds of animals mean different things. Zoos and museums are engines of difference, classifying and presenting the entangled mess of the natural world in a comprehensible way (ideally to be consumed in palatable chunks of time). Nature and culture are defined in particular ways in the modern West, and zoological collections are important sites for policing the boundaries between and within them. But the meanings afforded to particular species or genera by formal taxonomy belie the affective connections we make to, say, dogs (as Rachel Poliquin discusses in her essay).[20] Furthermore, pragmatic ways of arranging (such as by size) are as evident as scientific classification. The whale and the elephant have much in common.

But in the galleries of natural history museums, the striking visual differences are not between taxa but rather between modes of preparation; not between a lion and a tiger but between the skeleton and skin. The stored study specimen that Everest considers elicits very different reactions than does the mounted specimen on display in the gallery. Which is why, as Patchett, Foster, and Lorimer reflect, the "deadness" of the remains of the hen harrier is so significant. The most complete animals are to be found submerged in preservative fluid, but these "wet" specimens are often considered insufficiently palatable for display, notable exceptions such as the Natural History Museum's Darwin Centre notwithstanding.[21] Furthermore, the appearance and experience of a specimen has a great deal to do with how effective the preservation was. A well-crafted mount in a plush diorama is one thing, but visibly poor stitching and bulbous glass eyes are the stuff of horror movies.

Some taxidermists deliberately eschew lifelike forms. Steve Baker has coined the term "botched taxidermy" for contemporary art that presents animals and/or their remains that seem somehow to have *gone wrong*.[22] Such work draws explicit attention to the relationship between human and nonhuman animal, and the impact of the former on the latter. This is especially true of the subtle creations and interventions of Kate Foster (a coauthor, with Patchett and Lorimer, of an essay in this volume) and Snæbjörnsdóttir/Wilson (discussed in several essays), whose use of museum specimens seeks to engage with the animals on their own terms.[23] All taxidermy is artistic and has an aesthetic appeal, however distinctive: but the intentions and audiences involved in the avowed use of dead animals for fine art involves particular values and quite different afterlives.

Clearly, then, the meanings of animals on display are imbued not only (if at all) by those with custody of them but also by their audience, and therefore vary according to the ways they are looked at. To understand the afterlives of animals, we therefore address here their consumption as well as their production. Henning notes in her essay that in industrialized society our distance from some species objectifies them: watching an animal in captivity, whether in a cage or a case, has been likened to voyeurism.[24] On display, they come to represent things beyond their own material bounds. Osteological and other remnants act as maps, synecdochal of the whole organism. And although we have chosen for the most part here to write about individual animals, clearly they are more than singular when we look at them. They are also metonymic for their species, redolent of places far away and times long ago (even if they are not as ancient as visitors assume Maharajah to be, as I mention in my essay).

Nevertheless, the individuality of the animals is the overriding theme of this volume. Of the varied extrascientific meanings that recur throughout, the most striking is that of the fame of that particular beast. Balto and the Thames Whale were newsworthy, valiant individuals who undertook heroic journeys over ice and upriver (the former successfully, the latter with tragic consequences). They are more than metonyms, postmortem celebrities whose renown stemmed from their activities and/or visibility in life. The queen's zebra and Chi-Chi the panda had media presences that any Hollywood A-lister would envy (even if their meanings were commonly adapted for satirical purposes). On a local level, Maharajah, Sir Roger, and Alfred were "much-loved" mascots and emblems, instilling considerable identification with, and loyalty to, their institutions. The geographer Jamie Lorimer observes how significant this "nonhuman charisma" is to biodiversity conservation initiatives (see also Henry Nicholls's essay), identifying "flagship species"; here we see the connections people make with flagship individuals.[25] They have affective, political, and even humorous meanings, and our studies reveal that they continue to be popular after death; their renown continuing to develop postmortem. Whether or not museums capitalize on them, the social histories of key animals in the museum become part of civic consciousness, as any curator unwise enough to seek to remove an iconic exhibit of any kind will attest. Oral traditions run parallel (and sometimes even contrary) to formal museum documentation.

One key indicator of the fame of an animal in life or death is the nickname with which many of the present herd were mantled. This is also a key marker of the rampant anthropomorphism evident in these pages and reflected upon by Swinney.[26] "Nonhuman beings" they may be (see Hannah Paddon's essay), but they are instilled with many of our characteristics. Tellingly, this volume demonstrates that animals continue to be anthropomorphized postmortem, that personhood endures (or is afforded anew) beyond biological death, and that even in our retellings we contribute to this ambiguity, slipping between "she," "he," and "it." They hold a mirror up to us as writers (see Michelle Henning's essay); these tales are as much about people as animals.

The mandrill, hen harrier, and trophy mounts, meanwhile, in contrast to the famous beasts with whom they rub shoulders in this volume, are afforded neither human characteristics nor pet names. Selected for their anti-celebrity, they are among millions of anonymous specimens languishing (or not) in storage. Beyond Patchett, Foster, and Lorimer's astute reflections, we do not attempt here to address the massed meanings of drawer upon drawer of stored objects, but we are nevertheless aware of the limits of the single biographical study. To fully understand the many meanings within natural history museums, we must acknowledge not only the singular but also the standard, not quality but also quantity.

## PARAMETERS

Even if one were to study only renowned animals with names and posthumous careers, one could fill many more volumes like this. There are any number of further examples of iconic animals whose remains feature in modern museums: Martha, the last of the American passenger pigeons, is now at the Smithsonian; the Natural History Museum at Tring displays the famous interwar racing greyhound Mick the Miller, while Guy the gorilla is in the museum in London; General Custer's horse Comanche found its way postmortem to the University of Kansas Natural History Museum; and Dolly the cloned sheep has taken pride of place in the National Museum of Scotland.[27] Some are so famous that their disassociated skin, bones, and hearts are to be found in different institutions, such as Jumbo the elephant (see my essay on Maharajah), and even in different countries, like the racehorse Phar Lap, whose skeleton is in New Zealand (where he was born in 1926), but whose heart and mounted hide are in separate museums in Australia.[28] And as Swinney indicates in the afterword to this book, this is to concentrate largely on mammals: occasional invertebrates can also gain celebrity status, and dinosaurs, after all, are perhaps now the most iconic animals in museums.[29]

These conspicuous absences notwithstanding, the essays here present a herd of animals from across the world, whose afterlives were spent in Europe and North America. Contributors take variously museological, historical, anthropological, curatorial, artistic, and geographical approaches to the stories of these animals. They range from autobiographical to analytical, and anywhere in between (so we make no apologies for the variations in style, from poetic to clinical). For the most part, we are concerned with material beasts; Neurath's whale is an exception to this rule, but Henning is not the only author to consider the conceptual animal (as, for example, the circulation of Chi-Chi images). Most essays address an individual; Marvin's trophy mounts are the exception to this rule, but he is not the only author to discuss the general trajectory of animals on their way to museums (as I seek to do for elephants).

Given the title of this introduction, it would have been tempting to have arranged the essays two by two (we have a pair of elephants and a pair of whales), but instead they are presented chronologically. This, we hope, will give the reader an overview of historical changes (and continuities) in attitudes to animals in life and death, from eighteenth-century zebras to twenty-first-century aoudads. The noticeable cluster in the decades around 1900 is no coincidence; that period represents the peak in acquisition and status of natural history museums—our method is limited, after all, to those animals whose remains survive.[30] But in this concentrated period as well as the full three centuries covered by the volume as a whole, the essays show the changing meanings animals accrued during their journeys to and within zoos and museums; how the political contexts of collecting affected their value; what work was enacted on them as they became specimens, and how such techniques impacted upon their reception and use; and how conservators, researchers, and visitors engaged with these objects—as datum, as artifact, as fetish, or as ethical problem. As a whole, it is our intention that *The Afterlives of Animals* demonstrates the particularities of museum objects that were once alive.

We hope also that these studies will provide food for thought for (other) museum curators. As many historical natural science collections face uncertain futures, together these essays are suggestive of innovative approaches, interests, and angles to take on the use of these iconic specimens (see, for example, Patchett, Foster, and Lorimer's concluding thoughts). As I was drafting this introduction, colleagues in the Manchester Museum were installing a new exhibit in the foyer. As well as information on the habitat and conservation of the fauna in question, the accompanying text reflected on its afterlife, which bears similarity to those of Maharajah and the mandrill:

> The Museum has many stuffed animals, made from the skin of a real animal stuffed with cotton, plaster and other materials, not to mention the spooky glass eyes. This is a male Indian Swamp Deer (*Rucervus duvaucelii*) that once lived at Belle Vue Zoo, Manchester. When it died accidentally in 1907, it was offered to the Manchester Museum. Harry Brazenor, a local taxidermist, prepared it for display.

> Most of the Museum's stuffed animals date back to 1860–1900, so it is not surprising that some of them show their age. Many of these animals would have been killed—not by, or for the Museum—but as trophies

and curiosities. The Museum is firmly committed to nature conservation and we hope that by displaying these animals visitors will gain an
awareness of the natural world.[31]

Detailing historical and ethical information as a matter of course reflects the
questions visitors ask. Complementing research into taxonomy and biodiversity, it is the conviction of the authors that social and cultural analyses have
the potential to engage audiences, especially for aging displays. Taxidermy can
be appreciated as art as well as science; dioramas as historical documents as
well as natural heritage.[32] For us, the fascinating stories of animals on display
include not only information about their nature but also their culture; not
only their lives but also their afterlives.

## NOTES

1. Jane Carruthers, "Chapman, Abel (1851–1929)," *Oxford Dictionary of National
Biography* (Oxford: Oxford University Press, 2006), www.oxforddnb.com/view/
article/64726. Some mounts remain on display as part of the new "Biowall" display.
The Hancock was not unique in this approach: the Hungarian Natural History
Museum in Budapest, for example, includes an ark as part of its displays. (See A. M.
Tynan, "Abel's Ark: A New Way with Old Heads," *Curator: The Museum Journal* 30
[1987]: 85.94. Thanks to Professor Peter Davis.)

2. Nigel Rothfels, introduction to *Representing Animals,* ed. Rothfels (Bloomington and Indianapolis: Indiana University Press, 2002), xi.

3. Arjun Appadurai, ed., *The Social Life of Things: Commodities in Cultural Perspective* (Cambridge: Cambridge University Press, 1986); Lorraine Daston, ed., *Biographies of Scientific Objects* (Chicago: University of Chicago Press, 2000); Amiria
Henare, Martin Holbraad, and Sari Wastell, eds., *Thinking through Things: Theorising Artefacts Ethnographically* (London: Routledge, 2006); Stephen Harold Riggins,
ed., *The Socialness of Things: Essays on the Socio-Semiotics of Objects* (Berlin: Mouton
de Gruyter, 1994).

4. Mark Blackwell, ed., *The Secret Life of Things: Animals, Objects, and It-
Narratives in Eighteenth-Century England* (Lewisburg: Bucknell University Press,
2007); Christopher Plumb, "Exotic Animals in Eighteenth-Century Britain" (Ph.D.
diss., University of Manchester, 2010); see, for example, William Bingley, *Animal
Biography; or, Authentic Anecdotes of the Lives, Manners, and Economy, of the Animal
Creation,* 3 vols. (London: Phillips, 1803).

5. Richard L. Jaeschke, "When Does History End?" in *Archaeological Conservation and Its Consequences,* ed. Ashok Roy and Perry Smith (London: International
Institute for Conservation of Historic and Artistic Works, 1996), 86, also cited in Irit

Narkiss, "'Is This Real?': Authenticity, Conservation and the Visitor Experience," in *Art, Conservation and Authenticities: Material, Concept, Context,* ed. Erma Hermens and Tina Fiske, 237–45 (London: Archetype, 2009).

6. On Noah as collector, see John Elsner and Roger Cardinal, introduction to *The Cultures of Collecting,* ed. Elsner and Cardinal (London: Reaktion, 1994); on zoos as arks, see Vicki Croke, *The Modern Ark: The Story of Zoos: Past, Present, and Future* (New York: Scribner, 1998); Kurt Koenigsberger, *The Novel and the Menagerie: Totality, Englishness, and Empire* (Columbus: Ohio State University Press, 2007); and Nigel Rothfels, *Savages and Beasts: The Birth of the Modern Zoo* (Baltimore: Johns Hopkins University Press, 2002); on museums as arks, see Geoffrey A. C. Ginn, "An Ark for England: Esoteric Heritage at J. S. M. Ward's Abbey Folk Park, 1934–1940," *Journal of the History of Collections* 22 (2010): 129–40; and Arthur MacGregor, *Curiosity and Enlightenment: Collectors and Collections from the Sixteenth to the Nineteenth Century* (New Haven: Yale University Press, 2007).

7. The literature on natural history in different media is of course extensive: important studies on film in particular include Akira Mizuta Lippit, *Electric Animal: Toward a Rhetoric of Wildlife* (Minneapolis: University of Minnesota Press, 2000); and Gregg Mitman, *Reel Nature: America's Romance with Wildlife on Film* (Cambridge: Harvard University Press, 1999).

8. Susan Stewart, *On Longing: Narratives of the Miniature, the Gigantic, the Souvenir, the Collection* (Durham, N.C.: Duke University Press, 1993).

9. James A. Secord, "Knowledge in Transit," *Isis* 95 (2004): 663; see also Diarmid Finnegan, "The Spatial Turn: Geographical Approaches in the History of Science," *Journal of the History of Biology* 41 (2008): 369–88.

10. On the history of zoos, see Eric Baratay and Elisabeth Hardouin-Fugier, *Zoo: A History of Zoological Gardens in the West,* trans. Oliver Welsh (London: Reaktion, 2002); R. J. Hoage and William A. Deiss, eds., *New Worlds, New Animals: From Menagerie to Zoological Park in the Nineteenth Century* (Baltimore: Johns Hopkins University Press, 1996); Harriet Ritvo, *The Animal Estate: The English and Other Creatures in the Victorian Age* (London: Penguin, 1987); and Rothfels, *Savages and Beasts.* For different perspectives on natural history museums, see Stephen T. Asma, *Stuffed Animals and Pickled Heads: The Culture and Evolution of Natural History Museums* (Oxford: Oxford University Press, 2001); Richard Fortey, *Dry Store Room No. 1: The Secret Life of the Natural History Museum* (London: Harper, 2008); John M. MacKenzie, *Museums and Empire: Natural History, Human Cultures and Colonial Identities* (Manchester: Manchester University Press, 2009); and Monique Scott, *Rethinking Evolution in the Museum: Envisioning African Origins* (London: Routledge, 2007).

11. Thoreau's diary, 29 April 1837, *The Writings of Henry David Thoreau,* 20 vols. (Boston: Houghton Mifflin, 1906), 7:464. My thanks to Phill Simms for this reference.

12. Benedict R. O'G. Anderson, *Imagined Communities: Reflections on the Origin and Spread of Nationalism* (London: Verso, 1983), 180. My thanks to Malcolm Chapman for drawing my attention to this quote.

13. There is a burgeoning cultural and historical literature on taxidermy, surveyed in Samuel J. M. M. Alberti, "Constructing Nature behind Glass," *Museum and Society* 6 (2008): 73–97. Recent and innovative analyses include Jane Desmond, "Postmortem Exhibitions: Taxidermied Animals and Plastinated Corpses in the Theaters of the Dead," *Configurations* 16 (2008): 347–77; and Pauline Wakeham, *Taxidermic Signs: Reconstructing Aboriginality* (Minneapolis: University of Minnesota Press, 2008); for a journalist's perspective, see Melissa Milgrom, *Still Life: Adventures in Taxidermy* (Boston and New York: Houghton Mifflin Harcourt, 2010).

14. John Cleese as Mr. Praline, in Cleese and Graham Chapman, "Pet Shop [The Dead Parrot Sketch]," in *Monty Python's Flying Circus: Just the Words* (London: Methuen, 1999), 105.

15. Nigel Rothfels, "The Eyes of Elephants: Changing Perceptions," *Tidsskrift for kulturforskning* 7 (2008): 39–50.

16. James Griesemer, "Modelling in the Museum: On the Role of Remnant Models in the Work of Joseph Grinell," *Biology and Philosophy* 5 (1990): 3–36.

17. Mark Alvey, "The Cinema as Taxidermy: Carl Akeley and the Preservative Obsession," *Framework: The Journal of Cinema and Media* 48 (2007): 41.

18. On craft and technique, see, for example, Samuel J. M. M. Alberti, *Nature and Culture: Objects, Disciplines and the Manchester Museum* (Manchester: Manchester University Press, 2009), chap. 5; Merle Patchett, "Tracking Tigers: Recovering the Embodied Practices of Taxidermy," *Historical Geography* 36 (2008): 98–122; and Robert M. Peck, "Alcohol and Arsenic, Pepper, and Pitch: Brief Histories of Preservation Techniques," in *Stuffing Birds, Pressing Plants, Shaping Knowledge: Natural History in North America, 1730–1860,* ed. Sue Ann Prince (Philadelphia: American Philosophical Society, 2003), 27–53.

19. The Animal Studies Group, *Killing Animals* (Urbana: University of Illinois Press, 2006). See especially John Berger, "Why Look at Animals?" in *About Looking* (New York: Pantheon, 1980), 1–28; Erica Fudge, *Animal* (London: Reaktion, 2002); Donna Haraway, *Primate Visions: Gender, Race and Nature in the World of Modern Science* (London: Routledge, 1989); Tim Ingold, ed., *What Is an Animal?* (London: Routledge, 1988); Linda Kalof and Brigitte Resl, eds., *A Cultural History of Animals,* 6 vols. (Oxford: Berg, 2007); Ritvo, *The Animal Estate;* and Rothfels, *Representing Animals.*

20. On the history of taxonomy, see, for example, Harriet Ritvo, *The Platypus and the Mermaid and Other Figments of the Classifying Imagination* (Cambridge and London: Harvard University Press, 1997).

21. Phil Rainbow and Roger J. Lincoln, *Specimens: The Spirit of Zoology* (London: Natural History Museum, 2003).

22. Giovanni Aloi, ed., "Botched Taxidermy," *Antennae* 7 (2008), www.antennae

.org.uk; Steve Baker, *The Postmodern Animal* (London: Reaktion, 2000); Baker, "'You Kill Things to Look at Them': Animal Death in Contemporary Art," in *Killing Animals,* by The Animal Studies Group, 69–98; Libby Robin, "Dead Museum Animals: Natural Order and Cultural Chaos," *reCollections* 4, no. 2 (2009), http://recollections.nma.gov.au.

23. Patchett and Kate Foster, "Repair Work: Surfacing the Geographies of Dead Animals," *Museum and Society* 6 (2008): 98–122; Bryndís Snæbjörnsdóttir and Mark Wilson, *Nanoq: Flatout and Bluesome: A Cultural Life of Polar Bears* (London: Black Dog, 2006).

24. Berger, "Why Look at Animals?"; Randy Malamud, *Reading Zoos: Representations of Animals and Captivity* (New York: New York University Press, 1998).

25. Jamie Lorimer, "Nonhuman Charisma," *Environment and Planning D: Society and Space* 24 (2007): 911–32; Nigel Rothfels, "Immersed with Animals," in *Representing Animals,* ed. Rothfels, 199–223.

26. See, for example, Lorraine Daston and Gregg Mitman, eds., *Thinking with Animals: New Perspectives on Anthropomorphism* (New York: Columbia University Press, 2005).

27. On Mick the Miller, see Mike Huggins, "Everybody's Going to the Dogs"?: The Middle Classes and Greyhound Racing in Britain between the Wars," *Journal of Sport History* 34 (2007): 97–120; and Michael Tanner, *The Legend of Mick the Miller* (Newbury, UK: Highdown, 2003). Thanks to Dr. Hanna Rose Shell for information on Martha the pigeon, and to Boyd Zenner for suggesting Comanche, on whom see Elizabeth A. Lawrence, *His Very Silence Speaks: Comanche—The Horse Who Survived Custer's Last Stand* (Detroit: Wayne State University Press, 1989).

28. Paul Chambers, *Jumbo: This Being the True Story of the Greatest Elephant in the World* (London: Deutsch, 2007). Thanks to Dr. Helen Rees Leahy and Professor Brian Wheeller for drawing my attention to Phar Lap. At the time of writing, the Museum of New Zealand Te Papa Tongarewa is preparing to lend the skeleton to Melbourne Museum (even though the National Museum of Australia in Canberra had previously been deemed his heart too fragile to be loaned to Te Papa) (see www.tepapa.govt.nz).

29. Geoffrey N. Swinney, "'Granny' (*c.* 1821–1887), 'A Zoological Celebrity,'" *Archives of Natural History* 34 (2007): 219–28; W. J. T. Mitchell, *The Last Dinosaur Book: The Life and Times of a Cultural Icon* (Chicago: University of Chicago Press, 1998).

30. Samuel J. M. M. Alberti, "The Status of Museums: Authority, Identity and Material Culture," in *Geographies of Nineteenth-Century Science,* ed. David N. Livingstone and Charles W. J. Withers (Chicago: University of Chicago Press, 2011); Robert E. Kohler, *All Creatures: Naturalists, Collectors, and Biodiversity, 1850–1950* (Princeton: Princeton University Press, 2006). For an example of a longer-durée afterlife, see Taika Dahlbom, "Matter of Fact: Biographies of Zoological Specimens," *Museum History Journal* 2 (2009): 51–72.

31. Henry McGhie and Peter Brown, exhibit label for *Rucervus duvaucelii,* Manchester Museum, March 2010.

32. Eirik Granqvist and Adrian Norris, "Working Group on the Art of Taxidermy and Its Cultural Heritage Importance," *ICOM Natural History Committee Newsletter* 15 (2006): 1–6; Norris, "The Intangible Roots of Our Tangible Heritage," in *Intangible Natural Heritage: New Perspectives on Natural Objects,* ed. Eric Dorfman (New York: Routledge, 2011); Stephen C. Quinn, *Windows on Nature: The Great Habitat Dioramas of the American Museum of Natural History* (New York: Abrams, 2006); Karen Wonders, *Habitat Dioramas: Illusions of Wilderness in Museums of Natural History* (Uppsala: Almqvist and Wiksell, 1993).

It may seem ludicrous to close this chapter after so touching a narrative, with an exhibition of animals, but biography is necessarily mixed, and we must take our transitions according to the order of time, without considering the description of the events. Among other presents which were made to Her Majesty, a female zebra attracted most notice and excited considerable amusement.—John Watkins, *Memoirs of Her Most Excellent Majesty Sophia Charlotte, Queen of Great Britain*

CHRISTOPHER PLUMB

# "The Queen's Ass"

## The Cultural Life of Queen Charlotte's Zebra in Georgian Britain

So John Watkins's (1786–1831) biography of Queen Charlotte, after a particularly unctuous and sugared account of the late queen's domestic happiness and patronage of charitable institutions, dithered around the matter of the "Queen's Ass." Few in Georgian Britain were as restrained and would have denied themselves a smirk at Watkins's fastidiousness.[1] Queen Charlotte was associated with two living zebra in her lifetime, and they became a significant public representation of her character and that of her son in British culture.[2] These two zebra had a material history as they shifted between exhibitory contexts as well as between life and death. In the "afterlife," too, these zebra became a source of humor and satire as they became Hanoverian mascots. If the zebra that grazed at Buckingham House were ripe fodder for satirical commentary, they were also good for thinking about Enlightenment "improvement." Moreover, as the queen's zebra moved between different sites and practitioners—to showmen, anatomists, museums, and anatomical collections—the cultural life or influence of the zebra was diffuse and enduring. The zebra is a good candidate for understanding the symbolic potential and cultural significance of exotic animals in the eighteenth century.[3] Exotic animals acquired particularly strong political symbolism in matters of monarchy, and these associations were generated and circulated by public representations that foregrounded humor, satire, sexuality, luxury, and fashion.

The Hanoverians had a familial association with striped equids before the arrival of Queen Charlotte's first zebra in 1762. Frederick Louis (1717–1751), the Prince of Wales, had kept both a male and female zebra at Kew in the late 1740s and early 1750s. These zebra were described with accompanying colored plates in George Edwards's *Gleanings of Natural History* (1758), both drawn "from the living animal" at Kew and a "stuffed skin" at the Royal College of Physicians, London.[4] Frederick's daughter-in-law Charlotte and spendthrift grandson, Prince George Augustus Frederick (1762–1830), would later become especially associated with the zebra in public life.

The charisma of the queen's zebra in Georgian Britain, though a "ludicrous" closing vignette in the first chapter of Watkins's biography, is at the very heart of this essay as a cultural biography of the zebra. This zebra narrative, like the monochrome appearance of a striped equid, is structured into two distinct "stripes": first, the charisma of the zebra in Georgian humor; and second, the treatment of the zebra in natural histories as a principal character in Enlightenment discourses concerned with dominion over nature and "improvement." These particular stripes are indeed permeable because culture is, of course, not stark and variegated; it was quite possible to think about the queen's zebra with both registers in mind. Unlike the animal afterlives discussed in other essays in this volume, this biography is not primarily one of a singular animal. Instead, it is an eighteenth-century zebra with a hybridized pedigree: a composite of two different female zebra associated with Queen Charlotte, a male and female zebra belonging to Lord Clive and later dissected by John Hunter, as well as a zebra stolen from the king of Spain by privateers. The first section of this essay is particularly concerned with Queen Charlotte's two zebra and their charismatic place in Georgian culture as the "Queen's Ass." The second half is a broader cultural history of the zebra in eighteenth-century Britain, utilizing the zebra hybrid breeding experiment of Lord Clive and the character of Queen Charlotte's zebra in natural histories as case studies in wider Enlightenment discussion on naturalization and improvement. The separate "stripes" were closely related in Georgian Britain as the charisma and celebrity of the queen's zebra intersected with political critique and contemplation on the manipulation and tractability of nature. Georgian readers and spectators were aware of the multifarious and simultaneously respectable and seditious cultural life of the zebra in their national (and particularly elite) culture. Georgian English dictionaries and pronunciation suggest, too, why the zebra—the "Queen's Ass"—should have occupied a special place in humor and satire from the 1760s. The charisma of the zebra is shown to have been indebted to the historical phonology of the word "ass" and the amusement to be had in semantic slippage and connotations.

This appeal is perhaps why talking about the "Queen's Ass" was so pleasurable and enduring. John Watkins's truncated and dignified biographical account of the association between his queen and her zebra was then unnecessarily contrived; everybody knew what he was talking about anyway. After all, many in polite society had been having a good laugh for several decades.

## LAUGHING AT THE "QUEEN'S ASS"

In September 1761, King George III married the seventeen-year-old Duchess Sophia Charlotte of Mecklenburg-Strelitz (1744–1818), with a belated wedding present arriving in July 1762 on the HMS *Terpsichore*. Sir Thomas Adams, the ship's captain, had intended to present a pair of zebra (male and female) to the new young queen, but only the female made it to England alive. This zebra became a celebrity, and the queen and her eldest son, George, were associated with the "Queen's Ass" (as the zebra, the queen, and the prince became known) for several decades after the zebra's arrival. The cultural meanings of the "Queen's Ass" were diffuse in Georgian satire, where the representation of the zebra became a symbol of royal interests in the broadest sense and more particularly of critiques in the self-fashioning and behaviors of the Hanoverians.[5] The cultural life of the "Queen's Ass" was dependent on the original physical presence of the queen's zebra in 1762, but it also acquired robust associations that enabled it to exist for years beyond the lifetime of the queen's first zebra.

The role of King George III and Queen Charlotte as collectors and patrons has been the subject of considerable scholarship, yet neither their zebra nor their animal collections generally have received much critical attention.[6] Contemporaries certainly recognized the intellectual and cultural weight of their monarch's libraries, galleries, and cabinets but in other respects were distinctly unimpressed with the tone of the British court. George and Charlotte's matrimonial bliss and sense of domesticity (they had fifteen children) with an emphasis on understated private family life could be construed at best as naïve and sentimental by much of elite society. George and Charlotte's court was not a glittering Versailles. Characterized by many at home and abroad as moralizing, parsimonious, and, worst of all, unfashionable, the court was often ruthlessly lampooned in private letters and public satire. King George with his plain clothes, love of cold rooms, and preference for beef made a jocular "John Bull" or "Farmer George" to his promiscuous and profligate eldest son. Associated with happy domesticity, Queen Charlotte was undeserving of much of the lewd humor and innuendo heaped liberally upon her, as it was on many other elite women. Happily married with children, she was pretty unlikely to offer a paramour a peek of "Her Majesty's Ass." As we shall see later, the satirical function of the zebra became less sexual and more representative of the hegemonic interests of the queen and Tories.

The fashionable aristocratic beau monde, and especially the Whigs, poured scorn on the tedious court levées which they were often obliged to attend, and those without favor at court hardly mourned their enforced absence, instead creating their own envied circle as arbiters of taste and holders of political power.[7] Instead, "society" revolved around West London's grand residences like Devonshire House and aristocratic squares like St. James's crammed with leading Whig families. Prince George was glad to escape his staid, penny-pinching father, and his residence, Carlton House, became late Georgian London's most fashionable address. His sexual and financial liabilities were the bread and butter of London's Regency caricaturists.

The zebra were a popular sight (fig. 1), and crowds flocked to Buckingham Gate to view them as a royal gratuity. The *London Magazine, or Gentleman's Monthly Intelligencer* and other society periodicals announced the arrival of the first zebra in 1762, to be seen feeding in a paddock near Buckingham House.[8] For the benefit of those unable to get a closer view, a painting was made for display in the Mews stables.[9] The Queen's Guard had other ideas about free admission, and the first public outcry to emerge in early 1764 about charges the Guard was imposing prompted the queen's regiment to issue strict orders: "Complaint having been made that the Sentinels on the Queen's Guard extort money from persons that come to see the Zebra and Elephant, the Field-Officer in waiting desires that Officers in that guard to give strict orders that such unbecoming practices may be prevented in the future."[10] This order did little to prevent extortion of spectators' purses, and in the following years a number of indignant newspaper writers observed:

> Some servants were turned away last year for extorting money to see the Elephant and Zebra, owing to our gracious Queen's condescension to indulge the people with the sight, without any expense; but their dismission has not deterred their successors, who have absolutely refused to show the Zebra without being paid for it.[11]

> Is it consistent with dignity and decorum, that the property of the Sovereign should be exposed for pecuniary considerations under the very eye of Majesty? Do their Majesties know of such *petit* practices? Or, is the above a low trick of those in office to satisfy some wretched dispensation? What must foreigners, who judge of the whole by the characteristics of the few, think of such sordid doings?[12]

> Those abusive Scotch fellows who make a show of these animals, never were authorised to impose upon the public, by extorting money for a view of them. . . . this certainly cannot be known to her Majesty, whose study is to do everything in her power to please and indulge her subjects; it is a pity, therefore, her goodness, affability, and generosity, should be disparaged by such paltry wretches.[13]

Here the robbing of public purses became a matter of honor and national pride. The patriotic abhorred offense to royal dignity while the skeptical insinuated that the three-penny visitor fees kept the queen's coffers full. Later in the 1780s, the financial implications of maintaining a zebra were turned into a critique of the luxurious indulgences of the young spendthrift Prince of Wales, George Augustus Frederick. Described as "The Queen's Ass," George was both associated with his mother's penchant for zebra and mocked for his own vanity. Furthermore, the prince (and many others in power) were frustrated with Queen Charlotte's ability to curtail her son's political power with the assistance of Tory cronies, as her husband was increasingly perceived to be unfit to rule. Charlotte was making an ass of her son. In 1788, it was her turn to be represented as an ass, or rather a zebra (ridden by the Tory prime minister William Pitt), weighed down by gold-filled saddlebags containing "the Spoils of India and Britain." In the drawing, she brays: "What are Childrens rights to Ambition —I will rule in spite of them if I can conceal things at Q (Kew)" (fig. 2).

FIG. 1. George Stubbs (1724–1806), *Zebra,* 1763. Oil on canvas, 40½ × 50¼ in. (102.9 × 127.6 cm). (Yale Center for British Art, Paul Mellon Collection, B1981.25.617)

Prince George's 1787 portrayal in a zebra-striped three-piece suit is comedic but also bitingly satirical, undermining his masculinity, Englishness, and aesthetic judgment (fig. 3). There was a shared consensus among Whigs and Tories "that decrying effeminacy and luxury was an effective political strategy in a culture that associated political virtue with masculine simplicity."[14] In the case of a three-piece suit, the choice of cut, color, and material all attested to the adage that the clothes made the man. This sartorial association between the prince and the zebra was clearly apposite, since again in 1787 the poet William Wallbeck's *Zebra and the Horse* jibed:

> A zebra insolent and proud,
> Kept in the King's Menagery,
> Vaunting, as oft he did aloud,
> "None had so fine a coat as he"

FIG. 2. Queen Charlotte represented as her zebra, ridden by Prime Minister William Pitt (1759–1806). Thomas Rowlandson, *The Queen's Ass Loaded with the Spoils of India and Britain,* 1788. Hand-colored etching. British Museum. (© The Trustees of the British Museum)

"True!"—(Says the Hackney of a Squire
Who chanced along that road to pass)
"Your gaudery we must admire—
But, still, we know you for an ASS."

Similarly, the zebra as an exotic commodity could fall foul of mercantilist critique; imported luxury goods impinged upon British industry and trade, weakening the nation's moral fiber. Within a month of her arrival in July 1762, the zebra was providing ample fodder for satirists. In August 1762, "FART-inando" the "ASS-trologer" published a humorous song entitled "The Asses of Great Britain," and a large body of rump-related humor quickly amassed. The zebra proliferated in satires of 1762 and 1763 that scored political points against the new young queen and laughs at her expense.[15] Bawdy humor lent itself to the unusual quadruped and inspired a rather laborious ten-verse allegorical song:

A sight such as this surely was never seen;
Who the deuce would not gaze at the A___ of a Q___?
What prospect so charming!—What scene can surpass?
The delicate sight of her M___'s A___?

Though squeamish old Prudes with Invective and Spleen,
May turn up their Noses, and censure the Q___n;
Crying out,—"Tis a Shame, that her Q___nship, alas
Should take such a Pride—in exposing her A___."[16]

Other satires employed the charisma of the queen's zebra to articulate political concerns over perceived undue personal influence of the queen with
the king or government ministers. Some manipulated the conspicuous exotic
commodity status of the zebra as a means of underscoring unnecessary par-
simony elsewhere. Begrudging the penny-pinching absence of fireworks or
celebrations following the Peace of Paris in February 1763, to mark the end of
the Seven Years' War, the *Plain Dealer* even proposed that the queen's zebra
should take part in a mock battle.[17]

Polite society aspired to suppress unseemly laughter as a hallmark of gentil-
ity and instead cultivate amiable humor as a social currency distinct from the
low idioms of the plebeian mass. But ribald humour flourished in the highest
ranks as drinking songs and caricatures were purchased and collected for later
connoisseurship over a decent postprandial port. Displayed in the windows of
print shops, caricatures passed into the hands of individuals who then would
often disseminate to people in their circles of acquaintance and friendship
those that particularly tickled them or deliciously captured the flavor of current affairs or scandal.[18] There was plenty of scope here for semantic horseplay
since the English words "ass" and "arse" had a profusion of meanings. "Ass"
and "arse" were sometimes homophones (depending on pronunciation), and
scatological or sexual humour surrounding the "Queen's Ass" evinces that the
two words and their connotations often coalesced. "Arse," of course, referred
to the buttocks, but "to hang an arse" was also to be sluggish and tardy. An
"ass" was a stupid or dull fellow with lazy inclinations, and a clumsy person
could also be said to have fallen-over "arse-versy." A "jack ass" was an idiot, and
a punning appellation for the office of justice of the peace was, predictably,
"Just-Ass." A "black arse" was a burnt-bottomed kettle or pot, and "Ask my
arse!" a feisty street retort. In masculine collegiate circles, Cambridge scholars
and students wore their "cover-arse-gown."[19]

FIG. 3. S. W. Fores, *The Queen's Ass,* 1787. Hand-colored etching. British Museum. (© The Trustees of the British Museum)

Foreigners learning English or native speakers seeking guidance in matters of pronunciation had a profusion of dictionaries to consult. Gentlemen orthoepists like the actor Thomas Sheridan (1719–1788) became increasingly concerned about "remedying" the linguistic "peculiarities" or "vulgarities" of their provincial readers' tongues.[20] Clear orthographies from this period give some indication of how "ass" was said by some people and proscribed by others. Around 1762, when the queen's first zebra arrived, "ass" was sounded with a short vowel sound on the "a" and an enunciated extended "ss" to sound almost like an aspirated hiss with a distinct abrupt stop.[21] This "hiss" was not to the pleasure of some ears and by the late eighteenth century had attracted criticism. One rival orthoepist described the "hissing sound" as "the most disagreeable and reproachful of all our sounds, and therefore should not be affectedly extended."[22] Others thought so, too, and the orthography of "ass" in pronunciation dictionaries changed to reflect the shortened "ss" with less extension.

Given that Sheridan's diction was not that of an isolated and idiosyncratic actor since his works were widely read, it seems probable that his pronunciation of "ass" with a mannered and aspirated hiss was that used in the speech of the elite and their emulators between (at least) the 1760s and 1780s. By the late eighteenth century, the "aspirated ass" was an endangered species as English speakers dropped some of the linguistic traits and cultural habits of earlier decades that attracted criticism as capricious, effete, and affected. In the decades when jokes about the "Queen's Ass" had been most worth telling, the "aspirated ass" was likely to have been chosen for their delivery. The mannered and exaggerated pronunciation probably added to the humor and ludic pleasure of double entendres. The humor of the "Queen's Ass" and the biographical life of the first zebra to be associated with Queen Charlotte are entwined in the correspondence of the Rev. William Mason to Horace Walpole in 1773. Both clergyman and aristocrat were aesthetes, and their correspondence was suffused with literary flourishes, accounts of foreign news and travel, and political commentaries. With some pleasure, Mason (in provincial Yorkshire) wrote to Walpole, masquerading an exquisite tidbit as a "dull" desultory communiqué:

> This dull place affords me no news except that her Majesty's zebra, who, according to the advertisement in our *York Courant* of this day, it seems was lately the property of Mr Pinchy and purchased by him of one of her domestics (although, as I rather suspect, given to him for the valuable consideration of his friendship) died the third day of April last at Long Billington near Newark. This advertisement further adds "that the proprietor has caused her *skin to be stuffed,* and that upon the whole the *outward structure* being so well executed, she is as *well* if not better *to be seen now* than *when alive,* as she was so *vicious* as not to suffer any stranger to come near her, and the curious may now have a *close inspection,* which could not be obtained before." She is at present exhibited at the Blue Boar in this city with an oriental tiger, a magnanimous lion, a miraculous porcupine, a beautiful leopard and a voracious panther, etc., etc.[23]

The queen's zebra had at some point been sold (or given as a favor) to a Mr. Christopher "Pinchy" Pinchbeck (1710–1783), a clockmaker and friend to George III, and found herself on tour as part of a traveling menagerie. She then died and was "stuffed" and placed on display at the Blue Boar Inn in York, a far cry from the regal grounds of Buckingham House. The zebra's demeanor and the crowds that clamored to see her had denied many a close inspection (hence the commissioning of a portrait in her stable), which they could now get since the zebra was better to be seen now than when alive. Mason finished his letter with a flourish, alluding to humor familiar to Walpole:

> Pray do you not think the fate of this animal truly pitiable? Who after having, as the advertisement says, "belonged to her Majesty full ten years," should not only be exposed to the close inspection of every stable boy in the kingdom, but her immoralities whilst alive thus severely stigmatized in a country newspaper. I should think this anecdote might furnish the author of *Heroic Epistles* with a series of moral reflections which might end with the following pathetic couplet: "Ah beauteous beast! Thy cruel fate evinces / How vain the ass that puts its trust in Princes!"[24]

With her "immoralities" stigmatized in a newspaper and the sexual undertones to the "close inspection" of stable boys, the caricatures of Queen Charlotte and her zebra converge so that neither could be referred to without allusion to the other.

The afterlife of Queen Charlotte's second zebra, acquired in the 1780s, was somewhat more elevated than the "Blue Boar" in York. Charlotte presented this second zebra to the Leverian Museum, where it was displayed alongside the elephant that had also previously belonged to the queen and lodged at Buckingham Gate. At the Leverian, the affiliation between the queen and her zebra persisted, with guides to the sights of London in the 1790s deeming this relationship worthy of note. The 1800 Leverian Museum guide for children self-consciously perpetuated this cultural association so familiar to those of an older generation: "A zebra or Wild Ass, such as was presented some years ago (Mamma told us) to Her Majesty Queen Charlotte, did not escape our attention."[25] In late 1780, another royal zebra arrived in London, taken alive as booty by English privateers from a Spanish ship. Intended for the menagerie of King Charles III of Spain, the zebra was, instead, exhibited at the Bell Inn in the Haymarket near the Opera House for a one-shilling fee. The following week, the zebra was exhibited at Mr. Astley's Riding School near Westminster Bridge and advertised for sale at a hefty four hundred guineas (£420).[26] Later, in the early 1780s at Bartholomew Fair, the taxidermist and menagerist

Thomas Hall exhibited his "fine collection of stuffed birds and beasts" including a zebra; perhaps it was Astley's zebra that had passed into his ownership. In any case, a "mob" at Smithfields Fair in either royalist or anti-royalist sentiment purloined the "stuffed" zebra as a symbolic "Queen's Ass" and "drew it around the fair" to Hall's dismay. He never exhibited at the fair again.[27] The zebra as a royal mascot (Spanish booty and a mob trophy) appeared again in the 1770s and 1780s as two naval warships were named after the animal: the first HMS *Zebra* was launched in 1777 but scuppered and was blown up in 1778 during the American Revolutionary War. The second HMS *Zebra* was launched in 1780.

The charismatic appeal of Queen Charlotte's two zebra was clearly such that several decades after they had been visible to spectators while living in London, their presence in Georgian cultural life was enduring. As indicated earlier, this zebra biography is one of two "stripes"—one of satire and humor, and the other of the discursive function of the zebra in notions of enlightened "improvement" and changeable nature. With particular reference to the queen's zebra and to the zebra belonging to the former governor of Bengal Lord Clive (1725–1774), this biography will now turn toward the significant cultural life of the zebra that ran parallel with this satirical representation.

### HYBRIDS AND CARRIAGES

The collection of the Royal College of Surgeons of England contains two (extant) anatomical preparations of Lord Clive's male zebra, prepared by or for the surgeon anatomist John Hunter (1728–1793) in 1774; one in particular, the iliac artery from the zebra's penis (RCSHC/953), is important.[28] Hunter (or an assistant) dissected a section of artery to show contraction and elasticity in the penis, while Clive carried out a very different experiment with zebra reproductive organs when his mare was living. Clive, it would seem, made no (successful) attempt to breed his male and female zebra with each other, assuming they were kept in his menagerie at the same time, but he was certainly successful, eventually, in producing a hybrid from his female zebra. Lord Clive's successful attempt in 1773 to crossbreed a zebra with an ass is a dramatic example of the desire to adapt and render serviceable the zebra. After initial rejection of a stallion, the "shy" mare accepted her new partner (an ass) after the "extraordinary expedient of painting another ass so as to resemble a zebra." A male foal born in December 1773 "resembled both parents" and was considered likely to "propagate this species" further.[29] However, after Lord Clive's death in 1774, his menagerie was sold off and the extraordinary foal's whereabouts became unknown to hopeful naturalists. The male zebra at the Versailles menagerie in 1761 had "disdained" the female asses presented to him, leading the naturalist Georges-Louis Leclerc, Comte de Buffon (1707–1788), to conclude that "this coldness could be attributed to no other cause than the disagreement of their natures; for this zebra was then four years of age, and was very lively in every other exercise."[30] Clive's successful breeding between an ass with painted stripes and a zebra suggested that perhaps this disagreement in nature was pliable after all. The malleable nature of the zebra—or rather, the hope that the zebra would one day be tractable—was a significant feature of writing about the animal in Georgian Britain, with recurrent themes of domestication, natural adaptation, and wildness.

Both of Queen Charlotte's zebra were remarked upon as difficult, with biting and kicking particularly evincing their "ungovernable behaviour."[31] Confinement and a diet containing flesh and tobacco clearly took their toll on the queen's zebra, as their keeper was obliged to inform spectators of potential dangers. Indeed, nicotine-fueled bad behavior and ennui probably contributed to the passing on of the queen's first zebra to Christopher Pinchbeck after ten years in her menagerie—though her replacement in the following decade proved equally intractable. This zebra was later moved to the menagerie at the Tower of London. Here, the "irritability of her disposition" was demonstrated in an incident where she grasped her keeper with her teeth, threw him upon the ground, and "would have probably sacrificed his life to her resentment" had he not made a hasty retreat.[32] The vicissitudes of capture, a sea voyage, and captivity were seen to bear out adaptability in the diet of the zebra as it would feed on tobacco, flesh, and any other food offered.[33] The "iron law of necessity" could change some of the habits and diet of the zebra, but it was also observed that in other matters of character, the zebra, even when kept in a menagerie, remained unalterable: "All attempts to tame this animal so as to render it serviceable have been hitherto fruitless. Wild and independent by nature it seems ill adapted to servitude and restraint," wrote William Bingley, and William Nicholson considered the zebra to be "so wild and vicious as to give little hope that this beautiful race of creatures will ever eventually be of great service to mankind. . . . Should the zebra ever be made safely and easily convertible to the purposes of the horse, an elegance and variety would be added to the luxuries of the great and the opulent."[34]

The Enlightenment dream of tame or domesticated zebra pulling the liveried carriages of the great and opulent around St. James's was never realized, but was not in itself a naïve expectation for contemporaries, especially since in the last decade of the century a naturalized colony of kangaroos belonging to the queen grazed on the banks of the Thames at Kew and on the estates of the gentry and aristocracy.[35] This economic and cultural potential of an Enlightenment "improved" zebra is particularly meaningful when understood within much broader moves toward naturalizing, "improving," and disseminating other species of plants and animals throughout the empire. The botanical gardens at Kew cultivated in hothouses and flowerbeds specimens that through the vast network of Joseph Banks (1743–1820) might find their way to foreign climes, the landscape gardens of the English gentry and aristocracy, the dinner plate, or the manufactory.[36] In the animal kingdom, Spanish merino sheep were introduced and crossed with other varieties to produce higher-quality wool. Banks, Hunter, and others were particularly involved with the work of the British Wool Society in "improving" livestock. The British Wool Society hoped that Hunter might be a pioneer in this selective breeding of a goat suited to British climes but with down suitable for "India shawls" and other textile production. With such assiduous work in progress, the naturalist Oliver Goldsmith was optimistic that industrious Britons would prevail: "It is however, most probable that the zebra, by time and assiduity, might be brought under subjection, for as it resembles the horse in form, it has indisputably a similitude of nature, and only requires the efforts of a skilful and industrious nation to be added to the number of our domestics."[37]

The biographical life of the zebra—her cultural consumption and representation in Georgian Britain—extended from caricatures and natural histories to aesthetics and theology. The "Queen's Ass" left hoofprints across Georgian intellectual life, sometimes to comedic effect. Less successful was the attempt by some theologians and Hebrew scholars to suggest that the striped ass belonging to the queen was akin to that which Jesus rode into Jerusalem. Thomas Osbourne's *A Dissertation on a Passage of Scripture* (1792) posited that the "wild ass of the wilderness" tamed by Christ in the desert was a zebra, "the handsomest of its kind" and "like that presented to our most gracious queen," to which a critical reviewer responded sharply: "We can see no very important end to be answered by making out this point, except it be to bring the queen's *Zebra* into fashion among the modern objects of superstitious idolatry."[38] As late-eighteenth-century naturalists envisioned the future domestication of the zebra, so it was imagined that it had been achieved by divine power in the biblical past. The biblical provenance of the "Queen's Ass" did not become canonical theology, yet elsewhere the zebra received attention in the two eighteenth-century works in the canon of Georgian aesthetic theory.

Although natural histories consistently portrayed the zebra as an attractive animal, it is clear that some arbiters of taste were unconvinced by such fulsome praise. Writing in the 1760s, Reverend William Gilpin (1724–1804), the early progenitor of Picturesque taste, saw curiosity rather than beauty in the zebra; "the tiger, panther, and other variegated animals have their beauty but the zebra, I think, is rather a curious, than a picturesque animal. Its streaked sides injure it both in point of colour and in the delineation of its form."[39] It is possible that Gilpin saw Queen Charlotte's first zebra when formulating his opinions on the relative beauty of variegated animals, so perhaps his verdict was individual rather then collective, but either way he was unimpressed. Richard Payne Knight (1751–1824) in his *Principles of Taste* (1805) acknowledged this shortcoming in Gilpin's dismissal of striped equine beauty since one could not properly judge if a zebra was beautiful relative to others of its species without seeing more of that kind. However, the discerning could readily judge the more beautiful of different kinds of animal: "I never saw but one zebra, and one rhinoceros; and yet I found no difficulty in pronouncing the one to be very beautiful, and the other very ugly; nor have I met with any person that did." This was hardly an unqualified victory for the zebra, however, since Knight reserved his highest appreciation for his pet water spaniel, who with his long, curly hair afforded more play and variety of light, "a still more beautiful animal than a zebra."[40] Few spectators of the queen's zebra would have evaluated her with this vocabulary and stringent aesthetic criteria in mind, but this notwithstanding, some did, as Picturesque theory became a robust and culturally significant way of talking about and looking at landscapes and objects in Georgian Britain. Although the "Queen's Ass" was best known in her satirical incarnation, the presence of the zebra in theological and aesthetic treatises is indicative of the extent to which this animal, principally because of its material presence in Britain and its association with the queen, permeated cultural life.

## CONCLUSION

This cultural biography of Queen Charlotte's zebra articulates the intertwined charisma and characters of Sophia Charlotte and her zebra in Georgian Britain. Their associations with one another endured throughout their own lives and into a rich afterlife as Britons living in the second half of the eighteenth century continued to think and laugh about these two celebrities. The human and zebra as distinct categories in some cases slip so that commentators and spectators are conflating queen and ass; in other cases, the material presence of the zebra is independent of royal charisma. The personality of the zebra as an individual emerges in natural histories as temperamental, intractable, liable to bite and kick, tobacco- and meat-consuming, yet beautiful nevertheless. The bodies of the queen's zebra do not exist, though they once did at Buckingham House, the Blue Boar, and the Leverian Museum. But in their life and afterlife, material traces were left in printed texts, satires, and oil paintings that exist in the present. These attest to a rich culture of spectatorship in late-eighteenth-century Britain embracing the menagerie, museum, and print shop. The biography of the zebra demonstrates the contingent nature of spectatorship and the place animals hold in national and historical cultures. As Georgian Britons, and especially the masculine elite, looked at or thought about the zebra, the "Queen's Ass" reigned supreme as a cultural motif. When crowds first flocked to Buckingham House in 1762 to catch a glimpse, at least some of them would have been fully cogent of the satirical "Queen's Ass" circulating in ballads and broadsheets. Indeed, this knowledge probably enriched the pleasure to be had from spectatorship if it didn't prompt the desire to see the zebra in the first instance. When John Hunter dissected and prepared body parts of Lord Clive's zebra a decade later, he did so with an awareness of these cultural associations.

The cultural persistence of the "Queen's Ass" independent of the "real" zebra and enduring for decades is telling of the depth to which meanings generated by living animals persist in the afterlife. The binding of the characters of the queen and her zebra in British culture speak of the manner in which animals both living and dead become entrenched in relationships of affect and association. It is this charisma and the mental world or perspective of the spectator, often transient and difficult to interpret historically, that provides the means for better understanding past engagements with the animals, once living, that now spend their afterlives in museum collections—or, indeed, objects that were held within museum collections in the past and are no longer extant.

In a letter to Jean-Jacques Rousseau, Voltaire warned him of the English news writers and their habit of keeping an exact register of actions and jests, warning him that he would be talked about "as they do the Queen's Zebra, the English love to amuse themselves with oddities of every kind but this pleasure never amounts to esteem."[41] Writing in 1766, a mere four years after the queen's zebra arrived in England, even Voltaire would have marveled (and remained baffled) at the pleasure and amusement the English could derive from their equine oddity fifty years later. Voltaire was wrong. The English esteemed the "Queen's Ass" immeasurably, even if Queen Charlotte's biographer John Watkins could scarcely bring himself to write about her.

NOTES

1. The chronological boundaries of the Georgian period normally extend between 1714 and 1837. For the purposes of this essay, I am concerned with the Georgian period 1760–1820; the reign of King George III, the Regency, and the lifetime of Queen Charlotte.

2. Dorcas MacClintock, "Queen Charlotte's Zebra," *Discovery* 23 (1992): 2–9. For natural historical works on the zebra as a species, see Dorcas MacClintock, *A Natural History of Zebras* (Scribner's Sons: New York, 1976); and David Barnaby, *Quaggas and Other Zebras* (Plymouth: Bassett, 1996).

3. The circulation of exotic animals as commodities and their political symbolism in eighteenth-century Britain is discussed further in Christopher Plumb, "Exotic Animals in Eighteenth-Century Britain" (Ph.D. diss., University of Manchester, 2010).

4. George Edwards, *Gleanings of Natural History* (London: Royal College of Physicians, 1758), 222–23.

5. Diana Donald, *The Age of Caricature: Satirical Prints in the Reign of George III* (New Haven: Yale University Press, 1996).

6. Clarissa Campbell Orr, ed., *Queenship in Britain, 1660–1837: Royal Patronage, Court Culture, and Dynastic Politics* (Manchester: Manchester University Press, 2002); Jane Roberts, ed., *George III and Queen Charlotte: Patronage, Collecting and Court Taste* (London: Royal Collections, 2004).

7. Leslie Mitchell, *The Whig World, 1760–1837* (London: Hambledon Press, 2006); Hannah Greig, "Leading the Fashion: The Material Culture of London's Beau Monde," in *Gender, Taste and Material Culture in Britain and North America*

*1700–1830,* ed. J. Styles and Amanda Vickery (New Haven: Yale University Press, 2007), 293–313.

8. "Some Account of the Zebra, or Painted African Ass, Lately Brought over and Presented to the Queen," *London Magazine or Gentleman's Intelligencer* 31 (July 1762): 347.

9. George Stubbs (1724–1806) painted Queen Charlotte's zebra "from life" in 1762–63 and exhibited the painting with the Society of Artists in 1763. It is plausible that the painting commissioned to hang in the stables at Buckingham Gate was either a copy of this painting by Stubbs or an imitation done in his style.

10. "Extracts from the Orders Given to the Third Regiment of Guards on the 9th Instant," *Lloyds Evening Post,* 13 April 1764.

11. *Gazetteer and New Daily Advertiser,* 27 June 1766.

12. *Middlesex Journal or Chronicle of Liberty,* 21 September 1769.

13. Ibid., 16 January 1770.

14. David Kuchta, *The Three-Piece Suit and Modern Masculinity: England 1550–1850* (Berkeley and Los Angeles: University of California Press, 2002), 97.

15. See, for example, the broadsheet satires: *With a Fool's Head at the Tail: The Other Side of the Zebray* (1762); *The Real Ass* (1762); *The King's Ass* (1762); and *Zebra Rescued, or a Bridle for the Ass* (1762) (broadsides 1868,0808.4200; 1868,0808.4201; 1868,0808.4199; and 1868,0808.4203, British Museum).

16. Verses I and II from H. Howard *The Queen's Ass: A New Humorous Allegorical Song* (London, 1763) (broadside 1868,0808.4198, British Museum).

17. *Lloyds Evening Post,* 1 December 1763; *British Chronicle for 1763,* 2–5 December, 539.

18. See especially the essay by Vic Gatrell, "Bums, Farts and Other Transgressions," in *City of Laughter: Sex and Satire in Eighteenth-Century London,* 178–209 (London: Atlantic Books, 2006). More specific to sexual practices and humor, see Julie Peakman, *Lascivious Bodies: A Sexual History of the Eighteenth Century* (London: Atlantic Books, 2004). A more general history of humor is provided in Jan Bremmer and Herman Roodenburg, eds., *A Cultural History of Humour: From Antiquity to the Present Day* (Cambridge: Polity Press, 1997).

19. Samuel Johnson, *A Dictionary of the English Language* (Dublin: Jones, 1768); Francis Grose, *A Classical Dictionary of the Vulgar Tongue* (London, 1785); Nathan Bailey, *A Universal Etymological English Dictionary* (London: Ware, 1751); Eric Partridge and Jacqueline Simpson, *The Routledge Dictionary of Historical Slang* (London: Routledge, 2000).

20. Joan Beal, *English Pronunciation in the Eighteenth Century* (Oxford: Oxford University Press, 2002).

21. Thomas Sheridan, *A General Dictionary of the English Language* (London: Dodsley, 1780); Sheridan, *A Course of Lectures on Elocution* (London: Strahan, 1762).

22. James Adams, *The Pronunciation of the English Language Vindicated from Anomaly and Caprice* (Edinburgh: Moir, 1799).

23. William Mason to Horace Walpole, 2 June 1773, in *Horace Walpole's Correspondence 1756–1799,* ed. W. S. Lewis, Warren Hunting Smith, and George L. Lam (New Haven: Yale University Press, 1955), 90–91, emphasis added.

24. Ibid., 91. With "Heroic Epistles," Mason referred to the epistles of Ovid and possibly also to the poet Alexander Pope known for his Greek translations and epistolary style.

25. *The School-Room Party, Out of Hours* (London: Low, 1800), 64.

26. *Public Advertiser,* 15 November 1780; *Morning Post and Daily Advertiser,* 20 November 1780.

27. William Hone, *The Every-Day Book* (London: Hone, 1825), 1246.

28. Lord Clive's zebra died in 1774 and was dissected by Hunter. The collection of the Royal College of Surgeons of England contains the following two anatomical preparations of *Equus zebra* related to John Hunter: duodenum (RCSH/822); and iliac artery (RCSH/953).

29. William Nicholson, *Journal of Natural Philosophy, Chemistry, and the Arts* (London: Robinson, 1799), 267.

30. George Louis Leclerc Buffon, *Natural History, General and Particular,* 3rd ed., trans. William Smellie (London: Cadell, 1791), 265–66. Louise Robbins has written a short biographical account of the search for a zebra for the menagerie at Versailles and the zebra's subsequent arrival in 1761 (see "The Zebra Quest" in Louise Robbins, *Elephant Slaves and Pampered Parrots: Exotic Animals in Eighteenth-Century Paris* [Baltimore: John Hopkins University Press, 2002], 52–60).

31. Samuel Ward, *A Modern System of Natural History* (London: Newberry, 1775), 81.

32. Oliver Goldsmith, *An Abridgement of Dr. Goldsmith's Natural History of Beasts and Birds* (London: Whittingham, 1807), 14.

33. William Bingley, *Animal Biography or Popular Zoology,* 3 vols. (London: Rivington, 1820), 1:141.

34. Ibid., 2:122; William Nicholson, *The British Encyclopaedia; or Dictionary of Arts and Sciences Comprising an Accurate and Popular View of the Present Improved State of Human Knowledge,* 6 vols. (London: Longman, 1809).

35. Christopher Plumb, "'In Fact, One Cannot See It without Laughing': The Spectacle of the Kangaroo in London, 1770–1830," *Museum History Journal* 3 (2010): 7–32.

36. John Gascoigne, *Science in the Service of Empire: Joseph Banks, the British State and the Uses of Science in the Age of Revolutions* (Cambridge: Cambridge University Press, 1998); Gascoigne, *Joseph Banks and the English Enlightenment: Useful Knowledge and Polite Culture* (Cambridge: Cambridge University Press, 2003); Patricia Fara, *Sex, Botany and Empire: The Story of Carl Linnaeus and Joseph Banks* (London: Icon Books, 2004).

37. Goldsmith, *Natural History of Beasts,* 14.

38. Thomas Osbourne, *A Dissertation on a Passage of Scripture Little Noticed*

(London: Evans, 1792), 55; *The Analytical Review or History of Literature* (London: Johnson, 1792), 89.

39. William Gilpin, *Three Essays: On Picturesque Beauty, Picturesque Travels, and Sketching Landscape* (London: Blamire, 1792), 37.

40. Richard Payne Knight, *An Analytical Inquiry into Principles of Taste,* 2nd ed. (London: Payne Knight, 1805), 85.

41. François Marie Arouet de Voltaire, *A Letter from Mr. Voltaire to M. Jean Jacques Rousseau* (London: Payne, 1766), 35.

SAMUEL J. M. M. ALBERTI

# Maharajah the Elephant's Journey

## From Nature to Culture

In April 2009, the Manchester Museum, part of the University of Manchester, opened a new gallery displaying the connections between the history of the city and its people with the museum itself. The centerpiece was neither a spinning machine nor a steam train, but rather the skeleton of an Asian elephant. This imposing specimen had until then been displayed in the natural history galleries in a distant part of the building. He was Maharajah, "the elephant who walked to Manchester" in 1872: veteran of a traveling menagerie, a gaudy auction, and a zoo while alive; and of two museums and three galleries thereafter. His journey—from life to death, flesh to bones, "him" to "it," nature to culture—draws attention to the historicity of animal remains, and especially to the stubborn charisma of famous individual beasts. His tale is illuminating not only because of the many comparisons with other renowned animals displayed in life and death, but also because of his particular associations with place and site. His travels were geographical, chronological, and conceptual—embodying "knowledge in transit"—and along the way he invoked powerful responses from the people who encountered him.[1]

By the time Maharajah embarked upon his voyage, pachyderms were a relatively common sight in Britain. Elephants had toured the country in the seventeenth century, and in the early 1800s reinforced links with India and the proliferation of circuses and menageries swelled their numbers.[2] Among the most famous was Chuny (Chunee), who was displayed in London at Covent Garden Theatre and then nearby at the Royal Grand National Menagerie at Exeter Exchange on the Strand (the so-called "Exeter 'Change"). In March 1826, however, he became so agitated and violent that an impromptu firing squad was called to execute him, a fate widely reported (and roundly condemned).[3] Chuny's reputation was later eclipsed by a Victorian celebrity who was to become one of the most famous animals in the world. An African elephant born within a year of Maharajah was brought to the Jardin des Plantes in Paris from French Sudan, and then transferred to London Zoo, where he became known as "Jumbo."[4] To the anguish of the children whose safety may have been jeopardized by his increasing temperamentality, in 1882 the zoo sold Jumbo to P. T. Barnum for his famous American circus, in whose service he perished in a train accident in 1885.

But allegorical elephants are as interesting as their material counterparts. For if one is interested in the full range of scientific, affective, and political meanings associated with particular species, then elephants are a good place to start, not only because of their associations with colonialism but also because of their prodigious size.[5] Susan Stewart has argued that while "the miniature represents closure, interiority, the domestic, and the overly cultural, the gigantic represents infinity, exteriority, the public, and the overly natural."[6] Leviathans and behemoths, whether elephants, whales or dinosaurs, were vast chunks of mobile landscape (or seascape) ripped asunder and transplanted to the Victorian city as part of a culture of the spectacular that privileged the enormous.

Indian elephants' great bulk, like India itself, was to the Victorians reassuring and profitable in alliance yet terrifying when in revolt. Calm, working elephants were docile and helpful, stoically providing heavy labor for the imperial-industrial machine; by the nineteenth century, they were regarded as regal and intelligent, embodying "gentle, wry moralizing."[7] Regrettably their great size, and particularly that of their tusks, also made them vulnerable to ivory-greedy poachers. As it had been during the outcry at Chuny's assassination, the elephant-as-victim narrative was evident at the height of the ivory trade (and so in recent decades in conservation debates). And yet the elephant was a strange wild-domestic hybrid. It could be tamed, but didn't breed in captivity, and the celebrated docility could give way to terrifying rage, especially when the bull elephant was in the particular kind of heat known as *musth*. Such strength deployed in violence rather than service could be devastating.

These apparently conflicting characteristics of serenity and agitation, as well as associations with colonialism and the appeal of the gigantic, are demonstrated in the history of Maharajah: first in life and then after death.

## MENAGERIE TO ZOO

The journey of this particular elephant—at the least the recoverable part thereof—began around 1868, when he was purchased by Wombwell's Royal Number One Menagerie (presumably from one of the colonial animal-catching merchants satisfying the exploding demand from zoos and circuses).[8] Standing around five feet tall at his shoulder, he was not yet fully grown at four years old. He traveled alongside a female, pulling wagons (like Sir Roger later would for Bostock's Travelling Menagerie, as we will see in the next essay), attracting audiences and even rivaling the popularity of the menagerie's celebrated lion Hannibal. Like Hannibal, he was exhibited by Lorenzo "Lion Tamer" Lawrence, but he was also cared for by George Percival, who drove the elephant wagon. It may have been Percival or Lawrence who dubbed him "Maharajah" during this period, presumably in reference to his provenance, the one nickname among several that stuck. Like other elephants, Maharajah was celebrated for his docility and wisdom.

The Number One was one of several touring shows that stemmed from a collection of living animals that George Wombwell first exhibited at Bartholomew Fair in 1804. Wombwell was not the first to tour the country displaying exotic fauna in this way, but his was the largest and most famous of these peripatetic institutions in nineteenth-century Britain. Upon Wombwell's death in 1850, three of his troupes were touring the country, One and Two with the prefix "Royal" in recognition of their sojourns in various castles. Number One was run by Wombwell's widow, Ann (née Morgan). The year after Maharajah's purchase, Ann died and left the menagerie as a going concern to her nephew Alexander Fairgrieve. Unlike his aunt and uncle, however, Fairgrieve had no intention of dying in post, and in 1872 he made preparations to retire in Edinburgh. When the caravans arrived there for the last time, he first sold the horses, and on 9–10 April, after a final spell on display, Fairgrieve arranged for the menagerie—animals, caravans, and all—to be auctioned off in the Waverley Market.

Maharajah was one of the major attractions, by this time well over seven feet tall with impressive twenty-inch tusks. The sale itself was a performance, as the animals were put through their paces by Lawrence to the sound of the

ballyhoo of Mr. Buist, the auctioneer. This was a spectacular show, with ac-
tors (human and animal), a stage, props, and a script; the catalogue in effect
acting as a ticket of entry. Not only did the auctioneer and the bidders contribute to the resolution of value of the articles on sale, but so did the audience, who included in their number such notables as the zoologist Alphonse
Milne-Edwards of the Jardin des Plantes as well as representatives of Bostock
and Wombwell's Menagerie (another spin-off) and of Charles Jamrach's renowned animal dealership. They participated in a social process that resolved
ambiguities of classification and value—a public forum in which the worth of
unique, ancient, or living things was established. Animal auctions had a long
history, including such renowned sales as that for the Exeter 'Change, sold by
the famous natural history auctioneer J. C. Stevens.[9]

Bidding came along nicely, especially after Buist mentioned the prospect of
elephant sausages should a butcher put in a bid. As expected, Maharajah was
the largest single lot, fetching £680. The winning bid was from James Jennison, one of the family that controlled the Belle Vue Gardens near Manchester.
Jennison was there expanding their zoological collection; he also purchased
a baboon, an antelope, and the pregnant lioness Victoria. Pleased with this
bounty, Jennison booked his prizes on the 10:05 AM North British Railway
Company express train to Manchester on 11 April. Maharajah, however, had
other ideas. Upon entering the horse box reserved for him, the elephant trumpeted loudly, ramming the front of the box before bursting backward through
the locked doors. Although he was still young, this may have been the result
of *musth;* or there are numerous accounts of otherwise sanguine pachyderms
running riot out of character. (The renowned "Elefantino," or the "Venetian
Elephant"—now in the zoological collections of Padova University—refused
to board a ship in Venice and was eventually incompetently slaughtered, for
example; and Jumbo likewise took considerable coaxing into the crate to
transport him from London to Barnum's custodianship.)[10] Once back on the
platform, however, Maharajah returned to his usual docile self. It is possible
that he had revolted at his keeper's behest—Lawrence, after all, was facing
redundancy. One might further infer Lawrence's involvement from their departure later that day by road to Manchester, a lucrative and press-worthy
trek. Echoing famous journeys of Clara the rhinoceros in eighteenth-century
Europe and Zarafa the giraffe across nineteenth-century France, they walked
some twenty miles per day (well under the top speed of an unhitched pachyderm), attracting crowds—and money—as they went.[11]

FIG. 1. A very British elephant. Heywood Hardy, *The Disputed Toll,* 1875. Oil on canvas, 35 × 55 in. (88 × 140 cm). Manchester Art Gallery 1991.58. (© Manchester City Galleries)

Even though elephants were familiar to the British population, the journey was eventful. One episode stands out, despite almost certainly being entirely apocryphal. Elephant and keeper are alleged to have been held up at a toll by a parsimonious gatekeeper, at which point Maharajah simply lifted the gate from its hinges and continued his perambulation. The lack of evidential basis and an outright contradiction in the *Manchester Guardian* did not prevent the London-based sporting painter Heywood Hardy from taking the tale as inspiration for his *Disputed Toll,* shown at the Royal Academy in 1875 (fig. 1).[12] Hardy was accustomed to more traditional British scenes (and fauna), and it showed: his elephant has the hint of bovine about its physiognomy, especially in the way it extends its head over the fence. Elephants were often treated alongside or *as* cattle and horses, after all, both in India and once they arrived in Britain. Hardy's was not an Asian elephant so much as a British elephant. Compounding the implausibility of the episode, he included the tusk bar that was only fitted *after* the elephant's arrival at Manchester. Such canards notwithstanding, the painting changed hands several times in the following century—at one point fetching six hundred guineas—and was eventually given to Gerald Iles, the zoological director of Belle Vue.[13] He took it with him when he emigrated to Canada, but eventually donated it to Manchester City Art Gallery in 1991, where it is now on display. Its reemergence, meanwhile, prompted a local resident to come forward with a cruder version of the painting, acquired by the Manchester Museum in 1969, which was displayed beside Maharajah's skeleton. Bone and image, material fact and visual fiction, combined to construct meaning in the gallery.

Toll gates or no, after ten days and two hundred miles, Maharajah and Lawrence approached Belle Vue with impeccable timing midafternoon on Saturday, 20 April. The Jennisons had been alerted in advance, and had already trumpeted in the local press: "The Fine Male Indian Elephant, Maharajah, purchased at the sale of Wombwell's Menagerie, at Edinburgh, will arrive at the Gardens this day at about 2 p.m., having travelled by road from Scotland, via Carlisle, Kendal, Lancaster, Preston and Bolton, to Manchester."[14] Belle Vue was one of the earliest stationary zoological gardens in Europe, founded in the 1830s by James Jennison's father, John (see also Sophie Everest's essay in this volume). Intended "for the gratification of the public," it boasted a boating lake, maze, and fairground.[15] Hoping to gain a commercial advantage over other, neighboring pleasure parks such as the Pomona Gardens in Trafford in the 1860s, the Jennisons set out to reinforce their zoological offer. They made Belle Vue famous not only for its grand spectacles (including dazzling reenactments and a "Colossal Picture") but also for its charismatic animals. Elephants took pride of place on both counts.

The first elephant had arrived in 1860 from Ceylon, surviving for seven years in the crowded "menagerie" building. As a replacement in 1869 the Jennisons bought a young female Asiatic, "Sally," for £500. Maharajah's arrival three years later necessitated the construction of a temporary glass-roofed elephant house, rescuing Sally from the crowded menagerie; from 1876, they were housed, along with the new hippopotamus and rhinoceros, in a permanent brick building overlooking one of the lakes.[16] The elephants were presented as a couple, like Jumbo and Alice were at Regent's Park, and Hans and Marguerite had been at the Jardin des Plantes. The pair at Belle Vue and the four in London were among around thirty elephants in British zoos in the late nineteenth century, a large proportion of the one hundred across Europe.[17] The Belle Vue animals were described, classified, and laid out in a way that echoed museums and other more formal collections of living and dead fauna, explicitly a "collection," complete with descriptions and donor lists.[18]

Nevertheless, gathered with the other large quadrupeds near the monkeys, the elephants formed part of a vernacular taxonomy based on their size and appeal.

Pachyderms were also especially prominent because they were not only seen, but also touched. Walking among the visitors, the cage not their only stage, Maharajah and many other zoos' elephants became favorites of the thousands of children who rode them. He and Sally adorned the cover of the zoo's guidebooks and were central to spectacular shows such as "Napoleon Crossing the Alps," and in the city's May Day and Whit parades. As John Jennison's grandson George, later the zoo superintendent, remembered:

> Of the many elephants Maharajah was the chief. It lived 10 years in the gardens where it rode thousands of children [sic] and gave a very fine performance, in addition to forming the central incident in the "Prince at Calcutta," the firework spectacle of 1876. That was my first conscious connection with elephants. The keeper took me one night on the show, The Prince at Calcutta. I was 4½ years old. I have a vivid memory of pulling on his hat as my only uniform, and of being told to hold tight when the elephant knelt down for the Prince to descend. I also remember Cousin Albert, six months my senior, and very jealous of my luck, saying that I should get well whacked and I thought so too, but as a matter of fact nothing happened. Maharajah and Sally were the recognized leaders in the May Day procession. In Manchester their loads were hams from Broadys in Shudehill, and we little chaps, dressed in firework overalls, sat on the top of the load on the return journey. A shining 6d. from Mr Broady is still a happy memory. Maharajah attended many of the great fetes for which the Gardens catered at that time. It died in its prime as the result of an accident. One of the planks on the bridge across the lake broke beneath its weight and the fall splintered a tusk so badly that it had to be sawn off, with a great loss of blood. Maharajah lived for some considerable time, but was never the same animal afterwards.[19]

LIFE TO DEATH

Maharajah succumbed to pneumonia at the relatively young age of eighteen in 1882, the year Jumbo was shipped to America. As far as we know, his end was calm, in stark contrast to the terrible drama of incompetent firing squads

(Chuny and the Venetian Elephant), gunsmiths (Sir Roger), and a train (Jumbo). An exception to this "long elephantine necrology," his death was not the violent end of public anguish or hunter's pride.[20]

We have already established that this is not the end of Maharajah's story —for an elephant corpse is almost as interesting and valuable as the living beast.[21] On several occasions, the earliest elephants to Britain met untimely deaths, and armed personnel had to restrain crowds eager to see (and remove) their remains. One could pay to see the dead Chuny even before he was dissected by the anatomists Joshua Brookes and William Clift (who had been on hand to advise the most vulnerable area to aim at); parts of the flesh were then allegedly eaten—possibly the source of Buist's reference to elephant sausages at the Wombwell auction—and the skull inspected by the phrenologist Johann Spurzheimin in a final act of anthropomorphism. Zoologists likewise removed the Venetian Elephant's rotting remains (finding five hundred bullets in his hide). Jumbo's corpse needed a police guard as it remained for some time on the ill-fated railway embankment.

At Belle Vue, the Jennisons purchased a young African elephant soon afterward, followed by a pair of Asiatics; Sally was to live until 1901. None of them was to capture the public imagination like Maharajah, who adorned the Belle Vue guidebook for some years afterward.[22] He would also continue to attract audiences long after his death in material as well as emblematic form. There is no record of the fate of Maharajah's flesh and skin—one can find accounts of defleshing and anatomizing elsewhere in this volume—but we know his bones somehow made the short journey around the refreshment rooms, between the leopards and the stables, and up the spiral stone staircase above the menagerie room where Sally had lived before the elephant house was built. There, in the Belle Vue "Museum of Natural History," the skeleton was cleaned, prepared, and mounted, probably by Mr. Craythorne and his son James, who cared for Belle Vue's snakes and lizards (see fig. 2). The Craythornes' reptilian charges shared space with the museum, which had been erected in the middle of century:

> The Museum of Natural History, chiefly stuffed specimens of animals, made a far larger appeal than one would imagine probable and the Management thought it would be still more remunerative (there was 1d. admission) if placed near the centre. This was done in 1850 and marked the

FIG. 2. "An outlined mockery or a noble monument?" Maharajah on display in the Belle Vue Museum of Natural History with James Craythorne, ca. 1890. (Chetham's Library, Manchester)

beginning of a great zoological development. The Museum 75 feet long, 25 feet wide and 21 feet high, was built on the site of the small cages on the right of the Hyde Road Entrance.[23]

At the time of Maharajah's death, it was the only publicly accessible natural history collection in Manchester: for although the renowned museum of the Manchester Natural History Society had been transferred in 1868 to Owens College, predecessor of the University of Manchester, there it languished in the attic until the university completed a new museum building in 1890.[24]

Although the elephant was embedded in a tangle of living and dead nature —overlooking a small aquarium and astride a taxidermied lion—he clearly stood out:

Here in a large gallery are [sic] a miscellaneous collection of Birds, Animals, Reptiles, &c., almost all of which have lived in the Gardens. At the end of the room is a Skeleton of the celebrated performing Elephant "Maharajah," late of Wombwells, which at the dispersal of that renowned collection in the year 1872, was purchased by MESSRS. JEN-

NISON, and lived ten years in the Gardens. Here, too, may be seen two or three Cases of Live Pythons or Boa Constrictors . . . and also a case containing live Marmozets, lively and interesting animals from tropical South America.[25]

Maharajah was the only named individual in the museum until he was joined by the remains of the short-lived "enormous Orang-Utang, or Wild Man of the Woods of Borneo, caged in the Lion House in the summer of 1899."[26]

Belle Vue's juxtaposition of living and dead was not uncommon in menageries and circuses. Chuny's skeleton was for some time displayed in the same Exeter 'Change cage he had occupied in life, and various Wombwell menageries occasionally exhibited the remains of their star attractions, as did the Zoological Society of London (see Henry Nicholls's essay in this volume). Immediately upon Jumbo's death, P. T. Barnum announced, "If I can't have Jumbo living, I'll have him dead, and Jumbo dead is worth a small herd of ordinary elephants."[27] Aided by six local butchers, the taxidermist Henry Ward dissected the remains and removed them to his workshop in New York, which included Carl Akeley among the staff.[28] Ward's returned the mounted hide and articulated skeleton to Barnum, who toured with them and Jumbo's onetime companion, Alice, purchased from London shortly after Jumbo's death. If these macabre bones among living flesh lacked formal labeling, showmen were on hand; Barnum rehired Matthew Scott to tour with Jumbo's remains. Likewise at Belle Vue, one local paper recommended:

> Before we settle down to tea, let us step for half an hour into the Museum. We suspect the proportion of visitors to the gardens who even enter the Museum is very small, and yet it is one of the most interesting sights within the grounds. [The galleries are] each lined on either side with glass cases, in which are exhibited the stuffed forms of animals which have closed their public career in the gardens. If the attendant is in a chatty humour, you may learn from him strange tales as to their history.[29]

Even better at extemporizing was, of course, Lorenzo Lawrence, who was to work at Belle Vue for forty years. The journalist Filson Young interviewed Lawrence about his time at the gardens at the turn of the century: Young listened to the story of Maharajah, and then went to see the bones. The visual and tangible experience of the skeleton combined with the firsthand story of the elephant's life prompted a powerful response:

> Maharajah died of consumption [*sic*] years ago at Belle Vue, but the place still resounds with his fame; the awe in which he was held still lingered in the voices that spoke of him. Mr. Lorenzo told me that I should find his skeleton in the museum . . . I went, as to a shrine. I regarded, not without emotion, the vast ruinous framework supported by iron girders in the posture of life; I passed my hand over the great bleached bones, and in fancy saw them clothed upon with a bulk of living flesh. It is a strange thought, thus to make of the dead his own monument; a strange event that this grim creature, ravished from his far-off home to be a show for multitudes, should even in death have the privacy of the grave denied to his bones. There he stands, an outlined mockery or a noble monument —which you will. But even when his own fame is dead (and these bones shall long outlast it) his huge ruins may commemorate the prowess of Mr Lorenzo, his good friend and brother.[30]

The biographies of Lawrence and Maharajah were entwined, vividly recaptured in a multisensory experience: the sound of the tale, the sight and touch of the bones. The skeleton was at once a spectacle and a testament.

ZOO TO MUSEUM

Maharajah's skeleton was to be displayed in the Belle Vue museum for nearly six decades. But by the interwar period, the dead collection warranted fewer publicity inches, attracted fewer visitors, and their lively reptilian roommates had been transferred to new accommodation. In 1933, Gerald Iles (of the family who took over Belle Vue from the Jennisons) was appointed zoo superintendent at the age of twenty-one. He set about modernizing the zoo, and closed the museum—the specimens "were quite well mounted but time had faded them"—with a view to installing a new aquarium.[31] Although his fishy plans did not materialize, the museum remained closed, albeit intact, until finally, in 1941, the collection was dismantled and dispersed.

Maharajah's skeleton was to find its way from the commercial to the formal part of the exhibitionary complex (if they can be distinguished) like other iconic animals detailed in this volume and elsewhere. In death, Chuny spent time at William Bullock's Egyptian Hall before transferring briefly to the

University of London and finally to the Hunterian Museum at the Royal College of Surgeons, to be destroyed by bombing during the Second World War. Jumbo's skin was also destroyed in the twentieth century by a fire a Tuft's College, which had adopted him as its mascot after Barnum donated the skeleton. His heart, meanwhile, had allegedly been sent by Henry Ward to Cornell University; his skeleton remains at the American Museum of Natural History in New York.

Belle Vue had a long history of selling and donating its dead animals to the Manchester Museum, the natural science collections of the University of Manchester (as Sophie Everest explores in her essay in this volume). A few months before, for example, Iles had donated a mandrill, and later in the 1940s, the museum bought an elephant calf that was mounted with skin, and the remains of the female Asiatic "Lil" (who was with Bostock's before Belle Vue) also followed Maharajah to the university.[32] Living animals in turn returned to Belle Vue later in the century when the museum's crocodiles outgrew their accommodation. Furthermore, Iles would have been familiar with the collection from his zoology studies at the university. It was therefore the museum's zoologist, Cecilia Mirèio Legge, whom Iles contacted with a view to selling key items including Maharajah. "I am glad to note that the Museum Committee are interested in some of our specimens", he wrote in May 1941, offering "the Elephant, Capybara and Ant-Eater. . . . We should be very glad if you would be prepared to make an offer for these three exhibits."[33] Upon settling on £30 for the elephant and £7 for the others, Iles confided: "I am very glad the skeleton is remaining in Manchester, it is a fine exhibition piece and will look well in your galleries. . . . This elephant was purchased from Wombwell's Menagerie for £680, so I think you are getting it very cheap!"[34] (Cheap it may have been, complete it was not: Iles kept the sawn-off ends of Maharajah's tusks, and the bar installed on them, later displaying them alongside *The Disputed Toll*.)

As other Manchester Museum specimens were being boxed up and sent to the countryside for safekeeping, in June 1941, the Belle Vue acquisitions arrived.[35] Maharajah's skeleton became accession A.1225. It was part of a universal collection, an international scientific specimen representing the subspecies *Elephas maximus indicus* (identified by the French Georges Cuvier in 1798); the species *Elephas maximus* (identified by Swedish Carolus Linnaeus in 1758, although he ascribed this to both Indian and African elephants); the family

Elephantidae (identified by Briton John Edward Gray in 1821); and the sub-order Proboscidea (established by German Johann Illiger in 1811). It joined a selection of elephant remains already in the collection, including teeth, tusks, and a skull on display.[36] These other fragments, however, were tucked away in a bay. Maharajah was placed proudly in the center of the gallery, facing the signature staircase of the architect Alfred Waterhouse, and echoing another giant, the sperm whale skeleton hanging from ceiling above. He thereby took the same dominant place as his counterparts at the British Museum (Natural History) in an even grander Waterhouse building.[37] He represented not only elephants, but all Indian fauna, even all mammals. His skeleton and that of the whale were the first objects to come into view as visitors turned into the gallery: they encapsulated nature as a whole. In successive Manchester Museum guidebooks for thirty years, "attention [was] called to the mounted skeleton of the Indian Elephant which stands in the central space."[38] Certainly visitors could not miss him. As in life, he was especially popular with children, whether as part of the extensive formal education schemes or otherwise.[39] Even within the textbook-style guide of the 1960s, the museum's zoologist reflected on tales of "the intelligence of these long-lived beasts."[40]

Maharajah's appeal extended to the staff. Roy Garner, conservator at the museum for over three decades, reflected this affection when selecting his favorite specimen: "On the zoology side, it has to be, probably, Maharajah. Because we've moved Maharajah down, he used to be in the centre of the gallery, and you used to be able to look down, and his spine was S-shaped. You could see that he was slowly keeling over, so eventually we decided to strip him down and re-build him, so that was quite an interesting thing. We've done a lot of work on Maharajah, so I've grown quite fond of him."[41] Like all museum objects, Maharajah bore conceptual, affective, and physical traces of work. Nevertheless, his star had begun to wane. From the late 1970s, he was no longer mentioned in the museum's handbooks (unlike other "notable" specimens such as the Bengal tiger, the sperm whale, and even a spider crab).[42] As a new refit of the mammal gallery opened in 1989, Maharajah was to be found no longer in the center of the gallery but in one of the bays.[43] There he languished in an obscurity he had not suffered since the close of the Belle Vue museum in the 1930s. Children now allegedly expressed their fascination by repeatedly removing his tail (as Tufts students did with Jumbo) — although accounts of a secret stash of replacements kept in the plinth proved unsubstantiated.

With Maharajah no longer central to the museum's construction of nature, however, more attention was paid to the elephant's cultural significance. *The Disputed Toll* (albeit the rougher version) was displayed next to the skeleton, and the local author and keen zoologist David Barnaby researched and published a spirited book entitled *The Elephant Who Walked to Manchester* in 1988.[44] The (rest of the) skeleton was reunited with his long-lost tusk ends and the bar that so visibly demonstrated his provenance (eventually accessioned as A.2312.54). As the Manchester Museum undertook a massive capital development at the turn of the century, the new zoologist, Henry McGhie, gave tours and delivered talks emphasizing not only the elephant's natural features but also his social history. To celebrate the reopening in 2003, the artists Sophie Tyrrell and David Brown led a group of children in constructing a life-size papier-mâché model of Maharajah.[45] Elsewhere, other elephants were also reawakening: the Venetian Elephant enjoyed a five-year restoration, and Sir Roger emerged from his box to find himself at the center of the renovated Kelvingrove (see Richard Sutcliffe, Mike Rutherford, and Jeanne Robinson's essay in this volume).[46]

Maharajah thereby made an ideal candidate when plans were drawn up for a new permanent local history exhibition intended to explore the connections between the museum's collection and the wider city (initially entitled "Our City," but eventually dubbed simply "The Manchester Gallery"). Belle Vue featured heavily in the exhibition as an example of a mode of escape from the industrializing city, and both curatorial and exhibition staff considered Maharajah to be an ideal demonstration of these links while simultaneously providing an eye-catching center to the displays.[47] In November 2008, a team of seven staff dismantled the skeleton (again) and moved the separated parts from the animal life gallery, through Egyptology and anthropology to the conservation lab. Most parts were then carefully positioned on foam on open benches, but the huge rib cage served its time in exile on a custom-built frame. After months of preparation, staff and external experts rearticulated the bones over two busy days in March 2009. Finally, Maharajah returned to public view at the opening of the exhibition on 2 April. On a platform double the size of the old plinth, imposing on a new, bespoke steel frame, once again the skeleton dominated its space (see fig. 3). New interpretation panels retold the backstory of this iconic specimen alongside a video of one of the museum's "Collective Conversations" illuminating the diverse meanings of the elephant.

FIG. 3. "An impresive skellington." Maharajah on display in the Manchester Museum's local history gallery, 2009. (Photograph by Stephen Devine; copyright the Manchester Museum, University of Manchester)

The new plinth invited haptic exploration, and so Maharajah's multisensory appeal endured in the new gallery as it had in two museums, a zoo, and a menagerie. The media involved had changed, however. The museum used visitor-response cards extensively, and returns included such simple messages such as, "The elephant is fab."[48] A short film of the skeleton promoting the new gallery on the video-sharing website YouTube prompted "Dragonlordmike" to demonstrate the vernacular appeal of the gigantic: "Its such an impresive skellington when you see it up close, definatly recomend going to see the museum if not just for this and stan the t-rex in the fossils area."[49] The museum's presence on the micro-blogging service Twitter elicited a characteristically abbreviated response. "Maharajah the elephant is a brilliant feature," tweeted "Erikaspiteri," "myself and kids love looking at him and love his story about how he got to mcr."[50] On the video displayed in the gallery itself, schoolchildren, as fascinated as ever, interrogated members of the Gorton Local History Society about the experience of visiting Belle Vue, and about elephant morphology: querying the whereabouts of the trunk and tusk tips, and why there were so many ribs.[51] They also demanded to know whether the skeleton was "real or plastic." In the wake of the museum's high-profile acquisition of a *Tyrannosaurus rex* cast, their question was timely, and at least one journalist was also confused on this point.[52] To leap to clarify Maharajah's authenticity, however, is to miss an interesting point. Visitors climbing the stairs to reach the exhibition are unclear what monster faces them; expecting dinosaurs, they often assume it is a mammoth.[53] The lack of trunk and forehead cavity are striking in an elephant skeleton: for an instant, we seem to be faced with a giant, prehistoric, quadruped Cyclops.

## REFLECTIONS

"The appearance of the gigantic within the context of the city must be linked as well to the creation of public *spectacle*," Susan Stewart tells us; "the appropriation of the gigantic out of the natural landscape and its placement within the urban milieu of market relations marks a transition from the ambivalent (productive and destructive) forces of the natural to the reproductive and productive forces of class societies."[54] And yet Maharajah's journey was not a simple passage from nature to culture. Zoo elephants, circus elephants, even wildlife film elephants but especially museum elephants are a blend of cultural and natural. Never one to be shackled, Maharajah carried with him his rich history, sometimes apocryphal, sometimes in word and sometimes in image. His historical value was variously stagnant and strikingly in focus. *He* was at times singular and social; *it* was at other times scientific and metonymic. Like other elephants, his size and docility (mostly while alive, more reliably postmortem) attracted emotive and sensory responses, especially from children, but also from hardened zookeepers, conservators, and journalists. Just as mahouts and other keepers bond with elephants in life, so, too, do curators in death. Indeed, after five years, I, too, experience a connection with the elephant that goes beyond professional association: not, as my wife suggested, the affinity between two lumbering giants, but rather the bond formed between biographer and subject.[55]

If not nature to culture, then, perhaps Maharajah's passage from life to death is more revealing. On the one hand, his demise was not such a distinct event horizon as we might assume. After a brief period of disassembly, he remained on display near the space he occupied in his last decade. The study of museum practices reveals that to articulate is to articulate; rebuilding such beasts imbues them with new purposes, meanings, and explanations. Maharajah then formed part of another purchase exchange between institutions (albeit for a smaller sum), and continued his journey through the "exhibitionary complex" in death as he had in life—sometimes explicitly entertaining, at other times avowedly educational.[56] In Belle Vue as he would be in the Manchester Museum, he was a representative elephant in a zoological collection. In our enthusiasm to analyze the killing of animals, we should remember to pay attention to the geographical, institutional, and financial relationships that continue to be enacted through them postmortem.

Finally, an obvious point bears flagging that such life-death continuities are markedly different in an osteological specimen—Dragonlordmike's "impresive skellington"—than a taxidermic mount. The latter presents the illusion of life while the former is so strikingly dead as to be mistaken for a long-extinct beast.[57] However redolent of its past life a skeleton may be, it is clearly, sometimes alarmingly, dead; juxtaposed with taxidermy and even living animals, it can be a gloomy reminder of mortality. When we look skeletons, they do not look back; the elephant's eye, so wise in life, is absent.[58]

NOTES

1. James A. Secord, "Knowledge in Transit," *Isis* 95 (2004): 654–72.

2. Christopher Plumb, "'Strange and Wonderful': Encountering the Elephant in Britain 1675–1830," *Journal for Eighteenth-Century Studies* 33 (2010); Sujit Sivasundaram, "Trading Knowledge: The East India Company's Elephants in India and Britain," *Historical Journal* 48 (2005): 27–63.

3. Richard D. Altick, *The Shows of London: A Panoramic History of Exhibitions, 1600–1862* (Cambridge: Belknap Press of Harvard University Press, 1978); Harriet Ritvo, *The Animal Estate: The English and Other Creatures in the Victorian Age* (London: Penguin, 1987).

4. Paul Chambers, *Jumbo: This Being the True Story of the Greatest Elephant in the World* (London: Deutsch, 2007).

5. On the political meanings of elephants, see Carlos Gómez-Centurión, "Treasures Fit for a King: King Charles III of Spain's Indian Elephants," *Journal of the History of Collections* 22 (2010): 29–44; Louise E. Robbins, *Elephant Slaves and Pampered Parrots: Exotic Animals in Eighteenth-Century Paris* (Baltimore: Johns Hopkins University Press, 2002); Nigel Rothfels, "Killing Elephants: Pathos and Prestige in the Nineteenth Century," in *Victorian Animal Dreams: Representations of Animals in Victorian Literature and Culture,* ed. Deborah Denenholz Morse and Martin Danahay (Aldershot: Ashgate, 2007), 53–64; Liv Emma Thorsen, "A Fatal Visit to Ven-

ice: The Transformation of an Indian Elephant," in *Investigating Human/Animal Relations in Science, Culture and Work,* ed. Tora Holmberg (Uppsala: Centrum för Genusvetenkap, 2009), 85–96; Dan Wylie, *Elephant* (London: Reaktion, 2008).

6. Susan Stewart, *On Longing: Narratives of the Miniature, the Gigantic, the Souvenir, the Collection* (Durham, N.C.: Duke University Press, 1993), 70.

7. Altick, *Shows of London,* 316; Tamara Ketabgian, "'Melancholy Mad Elephants': Affect and the Animal Machine in *Hard Times,*" *Victorian Studies* 45 (2003): 649–76; Charles Knight, *The Elephant, Principally Viewed in Relation to Man* (London: Knight, 1844); Nigel Rothfels, "Elephants, Ethics, and History," in *Elephants and Ethics: Toward a Morality of Coexistence,* ed. Christen Wemmer and Catherine Ann Christen (Baltimore: Johns Hopkins University Press, 2008), 101–19.

8. David Barnaby, ed., *The Log Book of Wombwell's Royal No. 1 Menagerie 1848–1871 as Retained by George Percival, Driver of the Elephant Waggon* (Manchester: Zoological Society of Greater Manchester, 1989); Barnaby, *The Elephant Who Walked to Manchester* (Plymouth: Basset, 1988); John L. Middlemiss, *A Zoo on Wheels: Bostock and Wombwell's Menagerie* (Burton-on-Trent: Dalebrook, 1987); Nigel Rothfels, *Savages and Beasts: The Birth of the Modern Zoo* (Baltimore: Johns Hopkins University Press, 2002).

9. J. M. Chalmers-Hunt, ed., *Natural History Auctions 1700–1972: A Register of Sales in the British Isles* (London: Sotheby Parke Bernet, 1976); Charles W. Smith, *Auctions: The Social Construction of Value* (New York: Free Press, 1989)

10. Chambers, *Jumbo;* Thorsen, "Fatal Visit."

11. Michael Allin, *Zarafa: A Giraffe's True Story, from Deep in Africa to the Heart of Paris* (New York: Walker, 1998); Glynis Ridley, *Clara's Grand Tour: Travels with a Rhinoceros in Eighteenth-Century Europe* (New York: Atlantic, 2004).

12. *Manchester Guardian,* 11 May 1872; Barnaby, *The Elephant Who Walked.* I am grateful to Professor Alessandro Tosi, Università di Pisa, for his interpretation.

13. Gerald Iles, *At Home in the Zoo* (London: Allen, 1960).

14. *Manchester Guardian,* 20 April 1872.

15. *Guide to the Belle Vue Zoological Gardens near Manchester with a Description of the Colossal Picture* (Manchester: Heywood, 1872), 2; Robert Nicholls, *The Belle Vue Story* (Manchester: Richardson, 1992). On the history of zoos, see Eric Baratay and Elisabeth Hardouin-Fugier, *Zoo: A History of Zoological Gardens in the West,* trans. Oliver Welsh (London: Reaktion, 2002); R. J. Hoage and William A. Deiss, eds., *New Worlds, New Animals: From Menagerie to Zoological Park in the Nineteenth Century* (Baltimore: Johns Hopkins University Press, 1996); Ritvo, *The Animal Estate;* and Rothfels, *Savages and Beasts.*

16. *Guide to the Zoological Gardens, Belle Vue, Manchester* (Manchester: Belle Vue, 1878).

17. Charles John Cornish, *Life at the Zoo* (London: Seeley, 1895).

18. *Official Guide to Belle Vue Gardens* (Manchester: Belle Vue, 1909); cf. Iles, *At Home in the Zoo.*

19. George Jennison, "John Jennison—Belle Vue—Relating the Making and

Growth of the Famous Zoological Gardens, Belle Vue, Manchester, and the History of Its Competitors: A Century of Lancashire Open-Air Amusements, 1825–1925 . . . ," typescript, Disley, 1929, Chetham's Library Belle Vue Collection, F.4.11, pp. 111–12. See also "The 'Picture' At Belle Vue," *City Jackdaw: A Humorous and Satirical Journal,* 16 June 1876, 293–94.

20. Rothfels, "Elephants, Ethics, and History," 111; Ritvo, *Animal Estate.*

21. For details of the deceased pachyderms that follow, see, respectively, Plumb, "Strange and Wonderful"; Altick, *Shows of London;* Ritvo, *Animal Estate;* Sivasundaram, "Trading Knowledge"; Thorsen, "Fatal Visit"; Chambers, *Jumbo,* and Andrew McClellan, "P. T. Barnum, Jumbo the Elephant, and Barnum Museum of the Natural History at Tufts University," *Journal of the History of Collections* 23 (2011).

22. *Guide to the Zoological Gardens, Belle Vue, Manchester* (Manchester: Belle Vue, 1884); *Guide to Belle Vue, Manchester* (Manchester: Heywood, 1895).

23. Jennison, "John Jennison," 106–7.

24. Samuel J. M. M. Alberti, *Nature and Culture: Objects, Disciplines and the Manchester Museum* (Manchester: Manchester University Press, 2009).

25. *Official Guide to the Zoological Gardens, Belle Vue, Manchester* (Manchester: Belle Vue, 1891), 21.

26. *Official Guide to Belle Vue Gardens* (Manchester: Belle Vue, 1900), 22.

27. *New York Times,* 17 September 1885, cited in Chambers, *Jumbo.*

28. Frederic A. Lucas, "Akeley as a Taxidermist: A Chapter in the History of Museum Methods," *Natural History* 27 (1927): 142–52.

29. "Paterfamilias," "A Quiet Day at Belle Vue," *City Jackdaw: A Humorous and Satirical Journal,* 7 July 1876.

30. Filson Young, "Some Beasts and Their Keepers; by One of Their Friends: II—His Brother's Keeper," *Manchester Guardian,* 11 October 1899. See also Barnaby, *The Elephant Who Walked;* Young, *Memory Harbour: Essays Chiefly in Description* (London: Richards, 1909).

31. Iles, *At Home in the Zoo,* 23; Nicholls, *Belle Vue Story; Times* (London), 4 September 2004.

32. A.1303 (juvenile). Lil's remains were transferred to the university's Department of Anatomy rather than the Manchester Museum.

33. Gerald Iles to Cecilia Mirèio Legge, 21 May 1941, Manchester Museum Zoology. ZAC/1/79/1.

34. Iles to Legge, 26 May 1941, ZAC/1/79/2.

35. Mammal Accession Register, 1889–2009, Manchester Museum Zoology; "Acquisitions March-June 1941," Manchester Museum Committee Minutes vol. 5, p. 188, Manchester Museum Central Archive; *Manchester Museum Annual Report for 1940–1941* (Manchester: Victoria University of Manchester, 1941).

36. William Evans Hoyle, *General Guide to the Contents of the Museum* (Manchester: Cornish, 1892); Walter M. Tattersall, *General Guide to the Collections in the Manchester Museum* (Manchester: Manchester University Press, 1915).

37. William Thomas Stearn, *The Natural History Museum at South Kensington:*

*A History of the British Museum (Natural History) 1753–1980* (London: Heinemann, 1981).

38. George H. Carpenter, *A Short Guide to the Manchester Museum,* ed. Roderick Urwick Sayce (Manchester: Manchester University Press, 1941), 8; see also *Guide to the Manchester Museum* (Stockport: Dean, 1970).

39. Manchester Education Committee, *The City Art Gallery and the Manchester Museum Schools Service* (Manchester: Manchester Education Department, 1957).

40. Edmund L. Seyd, *Mammals,* 2nd ed. (Manchester: Manchester Museum, 1963), 14.

41. Roy Garner, interview by the author, 24 November 2004, "Re-Collecting at the Manchester Museum," disc S3, track 13, Manchester Museum Central Archive.

42. *The Manchester Museum* (Derby: English Life, 1985).

43. University of Manchester, *Annual Report* (Manchester: Victoria University of Manchester, 1989).

44. Barnaby, *The Elephant Who Walked.*

45. Carl Palmer, "Manchester Museum—Elephant Exhibition," *Manchester Evening News,* 30 October 2003.

46. Thorsen, "Fatal Visit."

47. "Our City" Gallery Brief, 2 March 2009, Manchester Museum Exhibition Archive.

48. Anonymous visitor card, August 2009, courtesy of Manchester Museum Data Group.

49. "Maharajah on the Manchester Gallery," www.youtube.com/watch?v=8CK6 qK3cu_g.

50. http://twitter.com/McrMuseum, 15 October 2009; http://twitter.com/ Erikaspiteri.

51. Abbey Hey Primary School interview Florence Wallwork and Janet Wallwork, October 2008, "Collective Conversations," ed. Adam St Clair, Manchester Museum Media Library.

52. Matthew Hobbs, "Star Elephant to Go on Show," *Manchester Evening News,* 13 March 2009.

53. Anne Speed, Visitor Services Assistant, interview by the author, 9 February 2010, interview VSA1; Maxine Byrne, visitor services assistant, interview by the author, 10 March 2010, interview VSA5, "Re-Collecting at the Manchester Museum," Manchester Museum Central Archive. On the changing meanings of mammoth and elephant remains over time, see Taika Helola Dahlbom, "A Mammoth History: The Extraordinary Journey of Two Thighbones," *Endeavour* 31 (2007): 110–14; and Adrienne Mayor, *The First Fossil Hunters: Palaeontology in Greek and Roman Times* (Princeton: Princeton University Press, 2000).

54. Stewart, *On Longing,* 84, original emphasis.

55. Dr. Fay Bound Alberti, Queen Mary University of London, personal communication to the author, 30 October 2009.

56. Tony Bennett, "The Exhibitionary Complex," *New Formations* 4 (1988): 73–102.

57. See also Thorsen, "Fatal Visit." For a survey of the literature on the history, purposes, and impacts of taxidermy, see Samuel J. M. M. Alberti, "Constructing Nature behind Glass," *Museum and Society* 6 (2008): 73–97.

58. John Berger, *About Looking* (New York: Pantheon, 1980); Plumb, "Strange and Wonderful"; Nigel Rothfels, "The Eyes of Elephants: Changing Perceptions," *Tidsskrift for kulturforskning* 7 (2008): 39–50.

RICHARD SUTCLIFFE, MIKE RUTHERFORD,
AND JEANNE ROBINSON

# Sir Roger the Elephant·

A mounted, twenty-seven-year-old, male Asian elephant: *Elephas maximus* to the scientists; accession number 1900.170 to the museum professionals; but Sir Roger to those who know and love him. Sir Roger is the most iconic and well-loved animal in the Glasgow Museums collections. At nearly ten feet (three meters) high, his size makes him hard to ignore. He has been a focal conversation point in the Kelvingrove Museum galleries since they first opened, more than one hundred years ago. The museum archives, the local press, the autobiography of his one-time owner E. H. Bostock, and personal knowledge give many interesting insights into his varied and interesting personal history.

## SIR ROGER ALIVE

Most of the information about Sir Roger's life comes directly from the autobiography of Edward Henry Bostock (1858–1940), who, following in the family footsteps, ran numerous traveling menageries (1878–1920) and founded the Scottish Zoo and Variety Circus in Glasgow in 1897.[1] Bostock became a prominent figure in Glasgow, and over the years donated many specimens to Glasgow's Corporation Museum, including Sir Roger. We don't know anything about the first ten years of Sir Roger's life. The entry in the museum's register reads "South India," so it is likely that was where he originated.[2] According to Bostock, Sir Roger had been with his Travelling Menagerie for twelve years prior to being settled in to the zoo in time for the May 1897 opening.[3]

The menagerie visited Scotland in 1884 and 1886, stopping in virtually every Scottish town between Berwick and Thurso, including small villages well off the beaten track.[4] As Bostock had acquired him about 1885, Sir Roger may well have come to Scotland with the 1886 tour. Bostock described him as "a quiet, well-behaved animal, and with the menagerie used to pull a small wagon on the journeys from town to town"—a description very similar to that of Maharajah in Wombwell's Royal Number One Menagerie in the 1870s (see Samuel Alberti's essay in this volume).[5] Sir Roger was greatly admired by visitors to the zoo, and many small boys eagerly spent their pocket money on buns to feed him.[6] Most of his time was probably spent in a small enclosure, but along with another younger elephant and several camels and dromedaries, he was taken out of the zoo on the New City Road and led out through the populous Glasgow streets toward the country for a couple of hours' exercise twice a week.

In October 1900, then aged about twenty-seven years old, Sir Roger, developed *musth,* a condition common to all mature male elephants during the breeding cycle. It is caused by the flow of a secretion called temporin from the elephant's temporal gland. Often painful, the condition can lead to unpredictable bad temper and aggression. The *musth* made Sir Roger extremely dangerous to handle, and he made several attempts to attack the zoo staff looking after him, including his regular keeper, John Allen. On returning from one of his regular trips out of the zoo for exercise, Sir Roger made a savage attack on Allen, who might have been killed if several members of the staff had not rushed to his assistance and beaten off the elephant. Even so, Allen was hospitalized with a broken arm and several broken ribs. A series of other keepers were found to look after Sir Roger, but the elephant soon showed his dislike for each man put in charge of him. Bostock himself could not get close to him without Sir Roger making menacing lunges toward him.

The elephant was normally tied up with chains and rings around one hind leg and one foreleg, but Bostock became increasingly concerned that these might not hold and that the elephant might break loose. A more secure enclosure seemed the best solution, and Bostock came up with the design. It was to be built beside where Sir Roger was kept, and so all the work had to be done while he was out being exercised. The builder took measurements the next time Sir Roger left the zoo, and the following time, workmen cut holes through the sixteen-inch wall of the building in order to attach a strong cage, which had been prefabricated beforehand. The whole construction was completed in less than two-and-a-half hours.

On his return from his walk, Sir Roger was installed in his new accommodation and made a very determined attempt to crush George Gruby, who up to this time had been the only member of staff he would still tolerate. With Sir Roger no longer allowing anyone near him, the staff had to throw his food to him and put down his water when he wasn't looking to avoid having both the bucket and water hurled back at them. Used buckets, often smashed, had to be rescued from his enclosure when his attention was diverted. Cleaning out his enclosure became impossible, and the smell from the *musth* and the accumulating dung became extremely offensive. Sir Roger's menacing attitude to visitors was such a cause for concern that Bostock reluctantly decided to humanely destroy him. Various methods were suggested, including poisoning and strangling, but Bostock decided that shooting him would be best.

### SIR ROGER'S DEATH

Arrangements were made with a Glasgow gunsmith, William Horton (a sporting gun, rifle, and cartridge, fishing rod and tackle manufacturer, of 98 Buchanan Street), to bring an elephant gun, and for two or three soldiers with ordinary rifles to come to the zoo on Thursday, 6 December 1900.[7] Bostock wanted to avoid a wounded and even angrier elephant breaking free and wreaking havoc, but, ever the showman, he decided to invite the paying public to witness the event, as reported in the *Daily Record:* "The 'execution' will take place this forenoon, a higher charge being made for admission to all those who wish to witness it from 10.30 till noon. After 12 o'clock the establishment will be opened at the usual prices, with the dead elephant on view."[8]

Bostock knew that Sir Roger was fond of wet bran, and he exploited this weakness to facilitate the elephant's end. The bran was put out, and it was agreed that when Sir Roger first put his head out for it, the firing party were to study the position; on the second occasion, they were to take aim; and on the third, to fire. The following day's *Daily Record* described how about one hundred men and one solitary woman paid the special price of two shillings to witness the execution.[9] Horton was armed with an Express rifle loaded with a copper-nosed bullet, and beside him were Captain Villiers and two sergeants of the Royal Scots Fusiliers, armed with service rifles with dum-dum and the usual Mark IV bullets. When the bran was put down for the elephant as arranged, they took aim; when he lowered his trunk the third time, the order was given, and they all fired simultaneously. Sir Roger slid to the floor without a murmur. The volley may have been fatal, but further shots were fired, just to make sure. It was later found that there were ten punctures in the head, and a hole in the right hind quarters showed that one of the Mark IV bullets had passed clean through the animal. The Express bullet fired by Horton had hit the center of the head with almost mathematical precision, and had probably passed right through the brain.

According to Bostock, "The volley killed the elephant instantly, and the big, mighty animal sank on his legs and died without a move."[10] However, according to an account of events published in 1947, this may not have been the case. Wullie Lindsay, who was a schoolboy in 1900, claimed to have been present and that "there was no apparent effect from the first volley, but each of the party quickly fired again on both occasions simultaneously, upon which Sir Roger collapsed quite dead."[11] A very violent death, in contrast to the calm end of Maharajah due to pneumonia (see Alberti's essay).

Bostock kept the body on show on Friday and Saturday, and thousands flocked to see the dead elephant, no doubt bringing in more income for the zoo. On Sunday, Bostock engaged a number of "butchers" to skin the animal (firm of "W. C. Hodgkinson & Co. Ltd, Licenced Horse Slaughterers, of 150 Old Keppochhill Road and 323 Kennedy Street, Glasgow"),[12] and he presented the hide and skeleton to the Art Gallery in Kelvingrove Park.[13] The butchers' premises were situated about a half mile from the zoo, and they were well used to dealing with a variety of animals that died there, as well as the usual horses and other domesticated animals which they treated on a daily basis. However, it is unlikely they had ever had anything on this scale before. It is unknown what happened to the elephant meat.

SIR ROGER DEAD

Dead elephants in the wild no doubt attract lots of vultures. In the case of a dead Sir Roger, it was professional people who moved in, wanting to take advantage of his body. The taxidermy firm of Rowland Ward of Piccadilly, London, wrote to James Paton, the superintendent of the museum, the same day that Sir Roger was shot:

Dear Sir,

We hear from Mr. Bostock that he has handed you an elephant. In case you wish the animal modelled we should be glad to undertake the work for you. No doubt you have seen the specimen we did for the Museum of Science and Art, Edinburgh.

> Yours faithfully,
> Per pro Rowland Ward Ltd.[14]

It is interesting that Rowland Ward's had apparently heard directly from Bostock. The story had been featured in the London papers, as well as in Glasgow, so it's possible they were implying a closer association with Bostock than they actually had in an attempt to curry favor.

A few days later (11 December 1900), Alexander MacPhail, recently appointed as a professor of anatomy at St. Mungo's College, Glasgow, wrote to Paton, asking if it would be possible for him to have part of the elephant's viscera for their anatomy museum:

Dear Mr Paton,

I notice that the mortal remains of the "Zoo" elephant have been bequeathed to the City, and as I am anxious not to miss any chance of improving one's knowledge of comparative anatomy I shall be interested to know if it would be possible for you to secure me any share in the animals [sic] dissection.[15]

Little had been offered in the way of a comprehensive description of the anatomy of the Indian elephant until the publication of an extensive paper by Louis Miall and F. Greenwood, curators in Leeds.[16] They pioneered a technique of preserving the organs out of fluid so that they could examine the whole carcass contents thoroughly over a three-year period. Subsequent publications after this date added small details, including features that were too damaged before dissection for Miall and Greenwood to describe, but by the time of Sir Roger's death, other anatomists were still keen to obtain material for study. MacPhail wrote again on 17 December, obviously rather too late, but giving the impression that he might still have a chance to rescue some parts for the college.[17] A copy of a telegram dated 29 December 1900 from MacPhail to John MacNaught Campbell (the curator) says, "If possible keep till Monday when shall be glad to receive material. MacPhail."[18] Unfortunately, having only one side of the correspondence, we do not know what the material in question was, and if it was anything from Sir Roger, or some other animal.

Around the same time, John Cleland (1835–1925) was a professor of anatomy at Glasgow University and was also building up a collection of anatomical specimens. There remain various elephant fragments in the Cleland collection, part of the anatomy collection of the university's Hunterian Museum: spleen, kidney, bile duct, stomach, tooth pulp, and an eyelid.[19] There is no indication as to where they came from in the first instance, but it is possible that Cleland got there before MacPhail, and that this elephant material is indeed from Sir Roger.

At Kelvingrove, (the remainder of) Sir Roger was duly added to the museum's register:

| '00–170 | Indian Elephant ♂ Elephas indicus, Cuv. South India. Shot in Bostock's Menagerie on 6th Dec. 1900. and had been in possession of donor for 16 years, and estimated to be 27 years old.[20] | E. H. Bostock "Sir Roger" | Scottish Zoo, New City Rd |

The museum's *Annual Report* for 1900 stated: "Mr E H Bostock still continues his generous gifts to the Natural History section of the Museum, his chief donation of the past year being a magnificent specimen of an Indian Elephant."[21] The work of mounting Sir Roger then fell to the taxidermy firm of Charles Kirk and Company. Kirk had been apprenticed at Rowland Ward's in London from about 1887 to 1894 and, after a couple of years in Perth, had moved to Glasgow in 1896, setting up his own business just down the road from Kelvingrove Museum at 156 Sauchiehall Street. From here he had been supplying high-quality taxidermy specimens to Glasgow, several other Scottish museums, the Bombay Natural History Society, and numerous private collectors.

There is only one known photograph of Sir Roger when he was alive. This was probably taken in the late 1890s or in 1900. The one really obvious feature is his lack of tusks. Although male Asian elephants usually have tusks, this is not always the case; those without are called *makhnas,* and are especially common among the Sri Lankan elephant population.[22] At some stage between his death and him being mounted for display, however, somebody decided that he ought to have some. Charles Kirk must have agreed to produce artificial ones, as a letter of 25 March 1901 reads:

> Dear Sir,
> Can you let me have per bearer the elephants tusk from which to model artificial ones for "Rodger" [*sic*]. I think it will be most suitable for a copy. I will let you have it back tomorrow as I'll take a mould in plaster. . . .
>
> Yours faithfully,<br>Charles Kirk.[23]

Kirk also requested the loan of "an elephant skull" to model (presumably Sir Roger's, as there is no record of another elephant skull in the collection).[24] It was loaned on 25 June 1901 and returned on 29 July 1901. This would have been required to make a lifelike manikin for Sir Roger's head. By the end of August 1901, Sir Roger had been mounted by Kirk, and he sent his account to the museum for payment on 2 September 1901, with the accompanying letter:

> Dear Sir,
> I enclose duplicate accounts for quarter ending Aug 31st. You will see I have included Hodgekinson's a/c for skinning and boneing [*sic*] elephant along with mounting, thought it better to do that, than make a separate item. The cost of mounting has exceeded my approximate estimate, but my expenses for material have turned out something like three times as heavy as I had expected; and the addition of tusks was an afterthought which entailed a considerable amount of work. . . .
>
> Yours faithfully,<br>Charles Kirk
>
> PS Elephant goes to "Zoo" on Wednesday first at noon.[25]

In the museum's accounts is an entry on 30 August 1901 detailing £85 paid to Charles Kirk for "Skinning, boneing [*sic*], preserving & modelling Adult Indian Elephant."[26] It would be interesting to know if the cost actually included an element for inconvenience, as apparently Kirk & Company had to remove their whole shop front in order to get Sir Roger out when he was finished.

Sir Roger is expertly mounted and reasonably lifelike, which has helped to ensure a long and engaging "afterlife" for the animal with his observers. The quality of the taxidermy makes a difference in the bond we can make with an animal. If it is badly mounted, it can be disturbing and repulsive, which is far from the reaction we see with Sir Roger. Maharajah the elephant had the tips of his tusks removed before he was sold to the Manchester Museum; they were of too much monetary value to donate in their entirety. In a way, missing parts seem to compromise the authenticity of an object less than the addition of parts that never were. Sir Roger is actually too complete with his fake tusks.

Once out of the shop, he didn't immediately go into his new home in the newly built Kelvingrove Museum. The building was in use as the Fine Art Section of the 1901 International Exhibition, held in Kelvingrove Park, which ran from 2 May until 9 November 1901. Instead, he went back to his former home at the zoo. Here he was put on display with the label, "'SIR ROGER' The Giant Elephant will be exhibited in the ZOO for a few weeks prior to being deposited in the Corporation Museum."[27] While he was there, he featured in a humorous article describing a Glaswegian family's outing to the zoo, including their discussion over his carcass about Sir Roger's fate. The father explains that he was shot and that he was dangerous, because he kept trampling his keepers and "eatin' the bunnets aff the folk's heids" (eating the hats off people's heads).[28]

## SIR ROGER IN THE MUSEUM

Despite an active life touring with the menagerie and being a popular exhibit in the Scottish Zoo, Sir Roger has undoubtedly touched more lives and wider audiences since his demise. The changes in the way that Sir Roger has been displayed and interpreted over the century reflect so clearly changing attitudes and relationship to animals and the natural world in society as a whole, and in museums in particular.

He must have moved to Kelvingrove Art Gallery and Museum at the end of 1901 or early 1902. The building opened to the public on 25 October 1902. It is not absolutely clear how he was displayed in the early part of the twentieth century. The new zoological exhibits in 1902 were to be a "careful classification on a scientific basis," so we can surmise that the mammals were displayed in taxonomic order in the east court of the building.[29] Sir Roger's skeleton had also been kept, and the bones were articulated in the museum's workshops in 1906 and put on display in the zoology section in Kelvingrove along with a large collection of other mammal skeletons in glazed cases.[30] At this time, the museum employed Edward Clinton Eggleton as an "articulator," and it is likely he did most of the work setting up the skeleton. By 1908, the east court was entirely devoted to mammals, and "the stuffed specimen and the skeleton have been placed side by side" (see Henry Nicholls's essay in this volume).[31] Wullie Lindsay recalled in 1947 that "there used to be an effect of an x-ray photograph, for, in another case, you can see his skeleton, with the hole in his forehead much larger than the one in his skin."[32]

On 13 and 14 March 1941, German air raids on Glasgow occurred, and land mines fell in Kelvingrove Park. The explosions broke most of the skylights in the roof, and tons of glass fell onto the exhibits below. Many of the natural history cases were badly damaged, and specimens also suffered (in some cases extensive) damage from the broken glass. The case holding Sir Roger's skeleton was one of those damaged, and so Sir Roger's skeleton was taken off display.[33] It remains partly disarticulated in storage. After the war, construction started on five new "animal group" dioramas in the natural history court. They consisted of African, Indian subcontinent, Australasian, Polar, and Scottish groups in very large cases built into the arches of the building. Each group was glazed slightly differently—the museum was using what was left of the large sheets of plate glass that had survived the wartime bombs.

Sir Roger found himself as part of the "Indian Habitat Group," alongside the young Asian elephant (received from Bostock in 1899), and surrounded by blackbuck, axis deer, red panda, Himalayan bear, tiger, and other Indian species. There was no interpretation in the display itself, other than an outline drawing of the diorama showing that he was an "Elephant, Adult male." However, a series of booklets were produced for sale in the shop. The *Animals of India* booklet explains that the case represents an Indian forest scene, and that "with a little imagination one can visualize the big elephant crashing through the undergrowth with the baby trotting along beside"; imagining him being led through the Glasgow streets or pulling a menagerie trailer wouldn't have worked so well in this situation. There is a general description of the Indian elephant, but no mention of Sir Roger. He is simply described as:

GENUS—Elephas   SPECIES—maximus
FAMILY—ELEPHANTIDAE

SUB-ORDER—PROBOSCIDAE
ORDER—UNGULATA.[34]

The old habitat groups weren't dismantled until 1981 to make way for the Natural History of Scotland Gallery. To allow for this major refurbishment, Sir Roger was banished to what was known as the "Zoo Room"—quite appropriate, considering his past. This was originally the Zoology Gallery, which was set up in the late 1950s, but closed in the late 1970s, and was temporarily being used as a storage space and also as a work area for volunteers mounting botanical specimens. Getting Sir Roger into the room was a close shave—almost literally. The arched doorway through which he had to go was exactly the same height as the mounted elephant. Sir Roger may have been slightly balder after his tight squeeze. It was not intended for Sir Roger to be off display for very long. The plan was for the "Zoo Room" to be reopened as a new gallery called Animals from Pole to Pole, featuring many of the specimens which had been removed from the other gallery. A sign promising the animals would be back in late 1982 was attached to the door of the gallery for some time, but this never happened. Things had moved on, and the budget didn't allow for it.

Sir Roger might have remained off display for a lot longer, but for the Glasgow Garden Festival. This was held on the banks of the River Clyde in Glasgow in 1988 and included an exhibition of fossil plants called "Time Trek." When the festival finished, "Time Trek" was offered to the museum. The only way it could be accommodated, however, was to reopen part of the "Zoo Room." The large mammals had to go somewhere else: Sir Roger was on the move again. And so, in January 1989, he was positioned in pride of place in the museum's center hall. His return was immediately celebrated in the local press, and he made the national news shortly afterward, when his picture appeared as part of a story on Museums Year.[35] After a few months, he was transferred back into the Natural History Court once again. This, however, was now the Natural History of Scotland Gallery, and he was an *Asian* elephant. It was therefore decided to tell his personal story as "Glasgow's Elephant"—just as Maharajah's social history was beginning to emerge in the Manchester Museum. This time, Sir Roger was prominently displayed with the young Asian elephant beside him as the centerpiece of the gallery. Remarkably, this may have been the first time that the museum had properly told Sir Roger's story—almost ninety years after he was first displayed. This put an end to some of the many myths which had grown up about him (more of which shortly). The label told Sir Roger's tale in Bostock's own words, quoting excerpts from his autobiography.[36]

Because of his popularity, some of the museum's staff employed Sir Roger to assist with fund-raising for the charity organization Comic Relief. A giant fiberglass red nose was made especially for him, and he wore it on Red Nose Day, the major biennial fund-raising day run by Comic Relief. In March 2001, he went one better. Comic Relief was saying "Pants to Poverty" (encouraging people to wear red underwear rather than the customary red noses)—and so was Sir Roger. He was dressed up in "a natty pair of long trunks" with a fifteen-foot waistband. The young elephant beside him sported a large diaper, complete with giant safety pin.[37] A donation box was placed beside them, and over a period of a couple of weeks, visitors donated a superb £1,253.17.

Kelvingrove Museum closed for a major refurbishment in June 2003. Almost all the objects in the museum were removed to a new purpose-built store at Glasgow Museums Resource Centre at Nitshill to allow the building to be gutted, refurbished, and redisplayed. Sir Roger was one of the three objects which were too big and heavy to remove from the building. A protective crate was built to house him, where he spent many months surrounded by scaffolding while the work went on around him. There was an inspection hatch on the side, and the museum's natural history conservation officer, Laurence Simmen, checked on him on a regular basis. Rumor has it that Simmen fed Sir Roger the occasional bun.

As Sir Roger had become such an icon of the museum in recent years, he was always going to be an important part of the new displays in Kelvingrove. The concept for Kelvingrove was that there would be lots of "story displays" throughout the building, grouped into gallery themes. The original plan was for a multidisciplinary display with objects relating to the zoo, menagerie, and elephants in general, but for financial reasons, a single disciplinary display was decided upon. Having led an international menagerie around the United Kingdom in life, it seems fitting that Sir Roger found himself at the head of a large procession of natural history specimens, which made up a story called "Nature's Record Breakers" within the Life Gallery (fig. 1). Sir Roger and the Baron of Buchlyvie, a famous Clydesdale horse, were the only animals to have their personal stories told. Despite being the largest specimen in Glasgow Museums' natural history collection, however, Sir Roger is not the largest object

FIG. 1. Sir Roger on display in the Life Gallery in Kelvingrove Museum in 2006. (© Culture and Sport Glasgow [Glasgow Museums])

in the building: just above and behind him is Spitfire LA198, which flew with the 602 (City of Glasgow) Squadron between 1947 and 1949. Because of its size, it could only be suspended in either the east or west courts, and so it appears to be just about to roar above Sir Roger's head. Sir Roger was key in the layout of the west court; each of the gallery entrances was to frame dramatic and iconic objects that would entice visitors in to the gallery space. As well as looking fantastic in his current position, Sir Roger is a useful, prominent, and readily identifiable landmark for directing people around the galleries. He also played a crucial role in the advertising campaign for the reopening of Kelvingrove. He featured on posters displayed all over Glasgow and appeared in various newspaper articles and other publications in the months leading up to the reopening in July 2006. When the museum did reopen, after being closed for three years, his popularity proved to be enduring.

A major part of the philosophy behind the new displays is that visitors encounter fewer physical and intellectual barriers. As a result, objects from all disciplines are now on open display in Kelvingrove. The designers had thought that psychological barriers would be enough to stop people from getting up close to the elephant. As tens of thousands of people poured in to see the new displays every day (more than 3 million visitors came to Kelvingrove in the first year after reopening), the temptation to lean over the psychological barriers and even lift their children onto the plinth so that they could touch Sir Roger proved insurmountable for many visitors. A low glass screen was later installed to protect him.

Sir Roger now features more prominently than ever in both external and internal media. He was filmed by Reef TV in March 2006 ahead of Kelvingrove's reopening, for a BBC2 series, *The People's Museum,* which allowed the public to vote for their favorite objects from around the country, which would then be added to a virtual museum.[38] Sir Roger wasn't part of the vote, but Mike Rutherford, one of the zoology curators, was interviewed to tell his story as a collection highlight. This resulted in a five-minute piece which featured in the Kelvingrove episode. More recently, Sir Roger also appeared on the BBC's *Bargain Hunt* (a television program which challenges contestants to buy antiques at a fair and then sell them in an auction for a profit) when the program visited Glasgow in 2009. He is featured in the Kelvingrove guidebooks and will be in a forthcoming mini-guide.[39] He is included in the museum's highlights tour and is used by the education team in a variety of ways. He has been drawn countless times during the annual schools art competition, and pictures of Sir Roger have won medals in the competition. An art student took accurate measurements of him in order to make his own reconstruction, but as far as we know, he has yet to find a gallery to take his proposed display.

## THE MYTHS

When his life was still fresh in the public memory, Sir Roger was in a taxonomic display with minimal labeling. His own story was not told in the museum until 1989. During the intervening years, various fanciful tales and urban myths developed around him. In 1947, for example, Wullie Lindsay claimed: "To this day, Jack Anthony tells how a toff [slang for upper-class person], in order to impress his girlfriend, pretended to offer Sir Roger food, but actually stuck a pin in his trunk. The toff returned a year later. Sir Roger remembered him and attacked him. Hence the reason for the shooting."[40] Four decades later, a local Glasgow newspaper connected Sir Roger to the story of an elephant shot, not at the zoo, but near Kelvin Bridge, with a grave dug in waiting for the skeleton.[41] Although the story is mostly speculation, it illustrates the level of public interest and highlights some ambiguous elements of Sir Roger's history. Why is there only one bullet hole visible in Sir Roger's head? Where are the others? And what happened to the original tusks?

Examination of the skull shows clear evidence of more than one bullet having done serious damage. There is one very obvious hole, measuring 1.3 × 1.6 in. (32 × 40 mm) at the top of the skull which lines up perfectly with an obvious indentation in the skin of Sir Roger's forehead. The hole passes right into the cranium and was undoubtedly caused by the bullet that killed him. There is also another hole, 0.7 inches (17mm) in diameter, lower down just off the center line of the skull. This has also penetrated the bone, but probably did not go right through to the brain. There is another large hole, through which a metal bolt has been placed, which secures the lower jaw to the skull. This may have been made by the articulator. Beside this there are two smaller holes, which may have been caused by ordinary rifles. There is another obvious hole just over half an inch (15mm) across on the right side of the skull just above the eye socket.

Why are there no signs on the skin of these other bullet holes? Well, in fact, there are. If you look carefully at his head, there are several smaller marks, which are almost certainly where the bullets passed through his skin. There are two of these on the left side of his head, fairly low down behind his mouth. On the right side of his head, there are more, just below and behind the eye. In addition, there is evidence of damage to the skin high up on the upper right hind leg—which would correspond with the bullet that was supposed to have passed right through him.[42] However, there is no clear evidence of Horton's other shot, which would have gone through the top of Sir Roger's trunk, on the skin. For Charles Kirk was a very good taxidermist: no doubt he and his colleagues simply filled in and patched the holes. For some reason, the position of the fatal shot is more obvious, perhaps just because it was a large deep hole, which was more difficult to hide.

Nearly a year later, the same journalist produced another story in the *Glasgow Guardian*.[43] He explored other versions of the legend: the elephant died of natural causes; there were two elephants, one male and one female, or possibly three; or even that the burial was somewhere else. The article then reiterated some of the details published about Sir Roger in a book, *The Good*

*Auld Days.*[44] However, he then added the postscript: "Mind you, this does not explain where the elephant's grave was (or is)." Perhaps all this grave speculation could have been kept at bay if the skeleton had been kept on display? The new interpretation of Sir Roger helps to dispel these myths and bring the story to a new audience. However, the oral history is often more powerful than the museum's interpretation. People often tell the story first, then check the label to see how well their memories are serving them.

In June 2006, the *Herald* (Glasgow) carried a story headed "Swing Low," which said the paper was trying to verify a story it had heard about a joiner sawing off the end of Sir Roger's trunk to make him fit on a plinth.[45] Myth? Legend? No—unfortunately, it was very close to the truth. All the new displays in the museum were built using modular structures. Sir Roger (on his original sycamore-wood base) was duly raised onto a standard plinth two foot five inches (735mm) above the floor. The idea was then to cover the original base (which over the years had been badly scuffed), with what was to be the standard Corian surface used throughout the gallery. Sections were to be cut to neatly fit around the four feet and trunk, but the external contractor who was doing the work decided it would make his job easier if he simply used a complete sheet of the Corian material at the front and cut the trunk instead. As the first reports of this happening filtered around the museum, everyone initially assumed it was a bad joke. To make matters worse, the contractor had thrown the sawn-off piece of trunk in the garbage. The museum's former natural history conservator, Dick Hendry, who was working on the installation of the new displays, did a very good repair using epoxy putty. Once painted, it was hard to see that anything had happened. But—like the elephant—we'll never forget.

What does the public think of Sir Roger today? Everybody seems to love him. Visitors make paintings of him, and many like to have their photographs taken with him. Numerous images of him have appeared on photo-sharing websites such as Flickr. One of the images has the following caption, which sums up Sir Roger in a nutshell: "This stuffed elephant was over 100 years old. Not when he died, but it was over 100 years ago when he was shot eating his breakfast because he'd become a little 'frisky' and was a danger to people. Poor Sir Roger."[46]

May his story and popularity live on for another century.

NOTES

1. Edward Henry Bostock, *Menageries, Circuses and Theatres* (London: Chapman and Hall, 1927), 161–65.

2. Glasgow Museums, Natural History Register, 1870–1903, 370, Glasgow Museums Resource Centre.

3. Bostock, *Menageries, Circuses and Theatres.*

4. George Eyre-Todd, *Who's Who in Glasgow in 1909* (Glasgow: Gowans and Gray, 1909).

5. Bostock, *Menageries, Circuses and Theatres.*

6. Wullie Lindsay, "I Saw Sir Roger Being Shot," *Evening Citizen* (Glasgow), 6 September 1947.

7. *Glasgow Post Office Directory 1901–1902* (Glasgow: Glasgow Post Office, 1901), 297.

8. "Condemned to Death—Big Elephant to Be Shot at the Zoo," *Daily Record* (Glasgow), 6 December 1900. Thanks to Roger Edwards for transcribing articles from the *Daily Record* in 2001.

9. "Elephant Shooting in the Scottish Zoo, Glasgow," *Daily Record* (Glasgow), 7 December 1900.

10. Bostock, *Menageries, Circuses and Theatres.*

11. Lindsay, "I Saw Sir Roger Being Shot."

12. *Glasgow Post Office Directory 1901–1902,* 293 and appendix 274.

13. Bostock, *Menageries, Circuses and Theatres.*

14. Rowland Ward to James Paton, 6 December 1900, Glasgow Museums Natural History Correspondence, 1900, no. 884.

15. Alex MacPhail to James Paton, 6 December 1900, Glasgow Museums Natural History Correspondence, 1900, no. 901.

16. Louis C. Miall and F. Greenwood, "Anatomy of the Indian Elephant," *Journal of Anatomy and Physiology* 12 (1878): 261–87.

17. Alex MacPhail to James Paton, 11 December 1900, Glasgow Museums Archives, GMA B0041/GMA204/922*.

18. Alex MacPhail to John MacNaught Campbell, telegram, 29 December 1900, Glasgow Museums Archives, GMA B0041/GMA202/950.

19. Maggie Reilly (Hunterian Museum, Glasgow) personal communication to Richard Sutcliffe, 23 October 2009.

20. Glasgow Museums, Natural History Register, 1870–1903, 370. It is now listed as 1900.170.

21. Corporation of Glasgow (Parks Department), *Museums and Art Galleries: Report for the Year 1900* (Glasgow: Corporation of Glasgow, 1901), 5.

22. H. I. E. Katugaha, Mangala de Silva, and Charles Santiapillai, "A Long-Term Study on the Dynamics of the Elephant (*Elephas maximus*) Population of Ruhuna National Park, Sri Lanka," *Biological Conservation* 89 (1999): 51–59.

23. Charles Kirk to Kelvingrove Museum, 25 March 1901, Glasgow Museums Natural History Correspondence, 1901, no. 140.

24. Loan form for elephant skull, 29 July 1901, Glasgow Museums Natural History Correspondence, 1901, no. 386½.

25. Charles Kirk to Kelvingrove Museum, 2 September 1901, Glasgow Museums Natural History Correspondence, 1901, no. 454.

26. Account Book, 1897–1902, 521, D-MAG/3, Museums and Galleries Department, Glasgow City Council, Strathclyde Regional Archives, Mitchell Library, Glasgow.

27. Black-and-white photograph of Sir Roger on display at the Scottish Zoo and Variety Circus, 1901, Glasgow Museums, accession number Z.1985.70.[2].

28. J. J. Bell, "A Visit to the Zoo," *Evening Times* (Glasgow), 1901, kindly transcribed by Roger Edwards in 2001.

29. Corporation of Glasgow (Parks Department), *Museums and Art Galleries: Report for the Year 1902* (Glasgow: Corporation of Glasgow, 1903), 5.

30. Corporation of Glasgow (Parks Department), *Museums and Art Galleries: Report for the Year 1906* (Glasgow: Corporation of Glasgow, 1907), 7.

31. Corporation of Glasgow (Parks Department), *Museums and Art Galleries: Report for the Year 1908* (Glasgow: Corporation of Glasgow, 1908), 8.

32. Lindsay, "I Saw Sir Roger Being Shot."

33. Ibid.

34. *Animals of India Habitat Group* (Glasgow: Corporation of the City of Glasgow, Art Gallery and Museum, Department of Natural History, 1947), 7–9.

35. "A Jumbo Draw Goes on Show," *Evening Times* (Glasgow), 30 January 1989; "Museums Year Clean Up," *Times* (London), 13 February 1989.

36. Bostock, *Menageries, Circuses and Theatres,*

37. "Trunk Call in Aid of Comic Relief," *Courier* (Bearsden & Milngavie), 8 March 2001.

38. BBC History, "The People's Museum," www.bbc.co.uk/history/programmes/peoplesmuseum, broadcast 23 June 2006.

39. Muriel Gray, *Kelvingrove Art Gallery and Museum: Glasgow's Portal to the World* (Glasgow: Glasgow Museums, 2006); *Kelvingrove Art Gallery and Museum— A Souvenir Guide* (Glasgow: Glasgow Museums, 2009); *Essential Kelvingrove* (Glasgow: Glasgow Museums Publishing and Philip Wilson, 2010).

40. Lindsay, "I Saw Sir Roger Being Shot."

41. M. Morton Hunter, "The Elephant Shot," *West End Times* (Glasgow), 31 May 1985.

42. "Elephant Shooting in the Scottish Zoo, Glasgow," *Daily Record* (Glasgow), 7 December 1900.

43. M. Morton Hunter, "The Elephant Remembered," *Glasgow Guardian,* 4 April 1986.

44. Gordon Irving, *The Good Auld Days: The Story of Scotland's Entertainers from Music Hall to Television* (London: Jupiter, 1977).

45. "Swing Low," *Herald* (Glasgow), 15 June 2006.

46. Tim Davis, "Sir Roger," www.flickr.com/photos/tmdbristol/2402860321.

SOPHIE EVEREST

# "Under the Skin"

## The Biography of a Manchester Mandrill

This essay explores the biographical potential of a male mandrill donated by Belle Vue Zoological Gardens to the Manchester Museum in 1909. During its short life, the mandrill traveled from the wilds of West Africa to the cages of a provincial zoo in the north of England. But for the last one hundred years it has been preserved as a study skin—its movement restricted to the various drawers, shelves, and rooms of museum storage. During the mandrill's journey through life and afterlife as commodity, zoo exhibit, and museum specimen, its meaning and status has been determined by cultural context. But unlike the biographies of Maharajah the elephant, Chi-Chi the panda, or Alfred the gorilla, there is nothing extraordinary about its history as either zoo captive or museum object. Instead, the animal's path through time and culture reveals a valuable typicality. As with the hen harrier in Patchett, Foster, and Lorimer's essay, the mandrill's current status as a study skin represents the majority of specimens that form our natural history collections today. These are objects whose trajectories rarely reach the gallery floor. Although they have been subject to the same processes of collection, curation, and conservation, access to this particular form of animal afterlife has been largely confined to the back rooms of museum storage and scientific research. But while the contemporary relevance of study skins as objects of scientific inquiry hangs in the balance, their potential as tools of cultural research emerges. Once resurfaced, their stories can be just as revealing.

PRE-HISTORY

The chartable biography of the Manchester Mandrill begins in the early 1900s, when the live animal was exported from Africa and sold to Belle Vue Zoo. But before the mandrill arrived on British shores, an idea of its species had long been established in the popular imagination. Mandrills had a presence in eighteenth- and nineteenth-century Britain, in the pages of popular natural history texts, and in the cages of menageries and zoological gardens. The earliest literary reference appears in William Smith's *A New Voyage to Guinea* (1744), in which he observes a "strange sort of Animal, called by the White Men in this Country, a Mandrill." The word "mandrill," Smith supposes, originates from its "near resemblance of a human creature."[1] In 1758, Linnaeus addressed this relationship of man to ape in scientific terms, in his definitive tenth edition of *Systema naturae*. Mandrills were classified for the first time as *Mandrillus sphinx* in the order of Primates in this edition. But Linnaeus's most notable addition to the order of Primates was *Homo sapiens*.[2] This action was the catalyst for much religious and scientific debate, and subsequent writers of natural history responded with a determination to distinguish man from beast. Buffon, Cuvier, William Bingley, J. G. Wood, and others emphasized the exalted position of man in a hierarchy of species. An animal's behavior and physical characteristics were described alongside judgments about its "nature" or morality, which corresponded largely to its resemblance or obedience to man. The garish coloration, four-legged gait, and doglike muzzle of the mandrill distinguished it from more humanlike primates, making it a particular focus for attack.

In 1817, Georges Cuvier found it hard to imagine "a more hideous or extraordinary animal" than the mandrill.[3] In 1829, William Bingley echoed, "it is difficult to figure to the mind an animal more disgusting in its manners, or more hideous in its appearance, than the Mandrill."[4] As Harriet Ritvo has argued, these judgments were more than an attempt to differentiate man from beast. The discourse of popular zoology became another means to define the proposed social order of the nineteenth century, and animals, like humans, were positioned accordingly. In the case of mandrills (as with other species), descriptions of physical form and behavior became implicitly bound to judgments about race, class, sexuality, and social decorum.[5]

In 1865, J. G. Wood described the mandrill's red nose, vivid posterior, and

THE MANDRILL.—*Papio Maimon.*

FIG. 1. Illustration of a mandrill from J. G. Wood, *The Illustrated Natural History: Mammalia* (London: Routledge, 1865), 75. (Wellcome Library, London)

cheek projections ridged with lines of blue, red, and purple as "repulsive," as if they were a deliberate assault upon Victorian taste and decency.[6] In the accompanying illustration, the mandrill greets the reader with a grotesque snarl (fig. 1). The indecency of the mandrill's physical form is matched by descriptions of the animal's sexual appetite as it becomes a metaphor for unrestrained human physicality and sexuality. At this juncture, the border between animal and human is further confused as the mandrill's sexual attentions are alleged to be directed most fervently toward humans. In *The Parlour Menagerie,* John Hogg writes that the male mandrill "frequently seizes and carries off women into the recesses of the forest."[7] These sexualized fantasies were not reserved for the male mandrill alone. In the early 1790s, a fantastical story circulated the press about a British sailor stranded on an island off the West African coast. It reports that he lived with a female mandrill for some years, fathering her three children. On his eventual rescue, the mandrill was sent into such wild despair that she threw herself and children into the sea to drown.[8] The tale resembles the nineteenth-century portrayal of the fallen woman, the inevitable climax of whose decline was to throw herself into the waters of the Thames.[9]

Alongside descriptions of sexual appetite and "fallen" behavior, criminality is a recurring theme. J. G. Wood depicts gangs of wild mandrills "plundering" entire villages of food while the male population was sent to work. This positioning of mandrill to native is revealing. The mandrills appear as a kind of parallel tribe, "formidable neighbours," "feared and hated by the inhabitants of Guinea." Wood emphasizes the "hopelessly savage" nature of the animal, in contrast to the native villagers, who obediently disperse to labor.[10] It is no coincidence that the mandrill's natural habitat along the West African coast was also the epicenter of the European slave trade. In these texts, mandrills represent a symbolic threat to the delicate social order established by European colonialism and must be subordinated accordingly. The animals encountered by white colonizers in West Africa and elsewhere were subject to the same politicizing forces as the native people. Within the discourses of popular zoology, the mandrill's unwillingness to submit to human influence and unbridled physicality and sexuality become neat metaphors for the stereotype of the uncolonized native.

While descriptions of mandrills in the wild dwelled on their savagery, a different representation emerged when the white man successfully exerted his control. Wood describes the technique used to capture and domesticate baboons with jars of beer placed near the animals' haunts. When the alcohol manifests its power, the baboons fall "easy victims to their captors."[11] In Wood's account, the craftiness of the animal catcher outwits the baboon as it transforms from savage beast to easy victim. The animal's undoing is its own dissolute morality and licentious appetite. The mandrill's appetite for alcohol is a recurring theme in descriptions of the captive animal in European menageries and zoos. The most famous nineteenth-century mandrill was "Happy Jerry," one-time resident of the menagerie at "Exeter 'Change."[12] Jerry was alleged to be partial to fermented liquors, and was routinely treated to a pot of porter, which he drank while seated in an armchair.[13] The use of alcohol to control the captive mandrill is echoed by the role of alcohol as an article of barter for human slaves in the colonial relationship. Just as the colonized African complied with his European master in the supposed subtext of popular zoology, the mandrill appears to be successfully contained within the confines of the European menagerie or zoo—willing to trade his freedom for a bribe of alcohol.

## THE MANDRILL AS COMMODITY

The parallel between mandrill and human slave resonates as the actual biography of the Manchester Mandrill begins with the long boat journey from West Africa through British docks. Exotic animals were a lucrative trade, and the ledgers of animal purchases kept by the Jennison family of Belle Vue Zoo in the early 1900s reveal an extensive network of international and provincial animal dealers. The precise acquisition route of the Manchester Mandrill can be traced within these pages to a handful of possible traders.

The mandrill was around six or seven years old when he died in December 1908 and was purchased by Belle Vue no earlier than 1901.[14] The zoo records show that sixteen mandrills were bought between the years 1901 and 1907. The vendors for these appear as William Cross, John D. Hamlyn, Mr. Wright, and Mr. Holmes Seward.[15] Both Cross and Hamlyn were major players and rivals in the exotic-animal business, a trade which had evolved rapidly since the 1860s in response to the growth of zoological gardens, circuses, and hunting parks throughout Europe. Dealers bought animals from incoming ships or via a network of agents in the colonies, reselling them through their own menageries in European port cities. William Cross was the grandson of Edward Cross, the famous proprietor of the Exeter 'Change. He operated from a number of premises in Liverpool, exploiting the city's long-established trade routes with West Africa. In 1901, the Cross menagerie sold three mandrills to Belle Vue at an average price of around £3 10s. each. Cross's name appears more frequently than any other alongside advertisements of mandrills for sale in the 1880s and 1890s. An advertisement posted by Cross in the *Era* newspaper in 1889 includes the claim: "Who imported more Large Baboons, Monster Blue-faced Mandrills, & the largest assortment of Monkeys, Apes &c than any man breathing? Cross, Liverpool."[16] The same newspaper includes a counterclaim by John Hamlyn, who boasts of supplying the Monkey Show in Regents Park and Belle Vue, among other leading zoos. Hamlyn operated from premises in St. George's Street in London's docklands. In 1907, four mandrills were purchased from John D. Hamlyn for around £4 per animal.[17] Both men are possible suppliers of the Manchester Mandrill.

The remaining Belle Vue mandrills purchased from Mr. Seward and Mr. Wright represent an alternative side of the market to Cross and Hamlyn. These men were using the same trade routes, but to compete with the larger firms, they sold animals to the zoo at a significantly reduced price. Holmes Seward sold four mandrills to Belle Vue between 1902 and 1905 at an average price of £2, and Mr. Wright also sold four mandrills to Belle Vue in 1906 and 1907, again at an average price of £2. It is interesting to consider the relative value of mandrills as a commodity during this period. While mandrills were purchased by Belle Vue in 1907 for as little as £2, in the same year a baby bear was bought for £5, a cape sea lion for £25, and an Indian elephant for £250. The largest purchase made by the zoo for this year was a rhinoceros at a staggering £1,000. Primates, on the whole, were bought for much less. Drill baboons average at just over £1 each, and twenty rhesus monkeys could be bought for £20.[18] In this context, the Manchester Mandrill occupied a routine or even mundane presence in early-twentieth-century Britain. Judging by the considerable number of mandrills for sale in the classified pages of newspapers such as the *Era* and *Daily News* in the late nineteenth century, there was no shortage of these primates arriving on European shores. Unlike larger exotic animals such as elephants or rhinos, mandrills were relatively easy to export. They regularly featured on the advertising bills of menageries, monkey shows, and zoological gardens. The mandrills listed in these newspapers and the ledger books of Belle Vue are not the hopelessly savage or deviant creatures of nineteenth-century zoological discourse but articles of trade that could be effectively shipped, processed, and sold. The mandrill objectified as a commodity was a manageable, conquerable entity.

## THE MANDRILL IN CAPTIVITY

Primates had been part of the collection at Belle Vue since 1845 and perhaps even earlier. In an advertisement for the zoo from 1850, special mention is made of the recently constructed monkey house: "a MONSTER CAGE, 21 Feet in Height and Covering an Area of upwards of 600 Square Feet."[19] In 1862, this cage was replaced by a larger "neat octagonal structure, glazed externally, with a space within enabling visitors to walk round the large wire cage."[20] By 1878, the primate collection had expanded to warrant additional cages inside the elephant house and crane enclosures, and in 1882, the construction of an elaborate new monkey house in the "style of the Indian Mosque or Temple" on the north side of the gardens.[21] This remained the main display area for monkeys at Belle Vue until the late 1930s and is highly likely to have been occupied by the Manchester Mandrill at some point in his life at the zoo. The building was kept to a temperature of 60–70 degrees Fahrenheit by a series of hot-water pipes; inside, ten small cages housed individual species. These regularly included a mandrill or pair of mandrills. Mandrills were also occasional inhabitants of the large central cage, which, according to the zoo's guidebook, contained a "grotesque and motley group of creatures." This cage was furnished with a variety of equipment, including a "Village Pump and Draw Well," an "Elevator" from which they collected water and corn, a "Rocking Horse," and a "Running Donkey."[22]

The architecture of display in the monkey house is particularly suggestive of the nineteenth- and early-twentieth-century relationship with the captive animal. Within the mock temple, the primates were reimagined in a fantasy of the civilized human world. The exterior shell of the building and mock-human furniture within the central cage anthropomorphized the animals for the entertainment of the viewing public while ultimately confirming their difference. An illustration in the 1895 guidebook shows visitors observing the monkeys from the main walkway. Men, women, and children point in wonder and lean in close. Monkeys return their gaze, clinging to the wire, and one looks to be stretching out its hand to a female visitor. Yet despite the suggestion of reciprocal looking, the subordinate position of the monkeys is confirmed by the wire partition between these fellow primates. Just as in the internal walkways of the nineteenth-century museum or national exhibition, the mechanism of display ultimately confirms the superior position of the spectator.

In a condensed version of the texts of popular zoology, the Belle Vue guidebooks regularly included descriptions of the animals housed in each cage. In the late nineteenth century, mandrills were described in the guide as "savage and destructive," conforming to the nineteenth-century stereotype of a hopelessly savage beast.[23] This view seems to be contradicted by a note on animal behavior and health in the zoo's logbooks from the early 1900s which describes the resident mandrills as "the most timid of monkeys."[24]

It is possible that the knowledge of and approach to this particular species was beginning to change by the time the Manchester Mandrill died late in 1908. There is evidence that mandrills and drill baboons enjoyed a particular status at Belle Vue as the pioneers for a new kind of open-air primate enclosure. A photograph from the early 1900s in the Belle Vue archive features a young mandrill and drill baboon reaching through the bars of their cage. Notes on the back of the photograph describe the animals as "growing well" due to the success of "open air treatment."[25] The inclusion of a colorfully illustrated mandrill on the cover of the 1909 guidebook—a position once occupied by Maharajah the elephant—is perhaps further evidence of the pride that the zoo felt in this particular exhibit at Belle Vue.

## DEATH AND PRESERVATION

Within the Manchester Museum zoology accession registers and correspondence archives lies the evidence of an intriguing relationship between Belle Vue Zoo and the museum. The Jennison family of Belle Vue were offering and donating the bodies of animals that died at the zoo to the museum's Zoology Department from the early 1900s. In return for these specimens, the zoo used the museum as a kind of consultant laboratory, asking for postmortem information on an animal's probable cause of death and advice on matters such as the infestation of insects on a plantation of willows or the discovery of a new species of earthworm.[26] Although no money changed hands between Belle Vue and the museum during this period, the dead bodies of animals operated as commodities offered in exchange for expert advice and information. The dead animals were the sites upon and through which a vital and mutually beneficial relationship occurred.

The Manchester Mandrill began its life at the museum via a letter written by James Jennison to the museum's director, William Hoyle, on 1 December 1908. Jennison informs Hoyle of the death that morning of a male mandrill baboon. He does not detail the animal's age or the cause of death, though the zoo ledgers reveal that a mandrill and other species of primates were dying from cold during the night that December.[27] In the letter, he writes that the mandrill is "a little finer than the ordinary specimens we have. Coming well into colour." Jennison offers the mandrill for the museum and asks Hoyle to call the following day to let him know if he is interested.[28] Exactly two months after Jennison's initial letter, on 1 February 1909, a mandrill appears as no. 597 in the museum's Mammal Accession Register. Under the field "Locality" is written "Belle Vue," and under "Remarks" is written "Messrs Jennison." The mandrill also appears in the list of donations in the museum's annual report for 1908–9.[29]

On entering the museum, the Manchester Mandrill was to undergo a radical process of physical change. Museum staff decided that the animal was to become a study skin as opposed to a taxidermy mount, and its trajectory from that moment took a particular course. To prevent deterioration and ensure its stability as a museum object, the mandrill first required preparation and preservation. Considering the zoo had approached the museum on the same morning as the animal's death, it is highly likely that responsibility for this task fell to museum staff rather than to Belle Vue or the museum's preferred taxidermist, Harry Brazenor. The museum reports for this period show that Assistant Keeper J. Ray Hardy was experienced in the preparation of zoological specimens for study and display.[30]

Whoever undertook this work, it is possible to reconstruct some elements of the process.[31] First, an incision was made using a knife or scalpel along the spine from the base of the mandrill's skull to the beginning of the tailbone. The skin was then peeled away from the flesh of the body and the skull, the spinal cord severed, and all flesh, organs, and bones removed from the abdomen and head. During this process, absorbent padding would have been used to keep the skin and hair clean from the inevitable amount of blood and mess. Salt was then rubbed into the skin to prevent drying. At the base of each limb, an incision was made sufficient to sever the bone and remove it by peeling the skin away. The underside of each digit was split open and all flesh scraped clean, leaving the smaller bones in place. The mandrill skin was then cleaned of fat with a knife or the back of a spoon and washed with soap. After the skin was dried with sawdust or a substance such as magnesium carbonate, the preserving process began in earnest. First, a salt mixture was rubbed into the skin, creating an antibacterial environment. In 1909, it is highly likely that the soap or salt mix contained arsenic, creating a poisonous layer which killed any bacteria or insects reaching the inside of the skin. After chemical preservation, the abdomen, head, and limbs of the mandrill were stuffed with a substance such as wood wool, flax fiber, or cotton wool and the skin sewn up along all incisions and orifices.

As a museum practice, it is important to distinguish this method of preparing a skin from that of creating a taxidermy mount. Lifelike representation is not a consideration in the preparation of a study skin, while mounted taxidermy typically involves the creation of a manikin in a lifelike pose, over which the skin is then repositioned. Although the manikin method was not employed until the early 1900s, this distinction dates back to the earliest years of recognized museum taxidermy, when naturalists debated the respective merits of study skins for scientific research and naturalistic, lifelike displays. The renowned naturalist William Swainson decried the artificiality of mounted displays in the early nineteenth century, insisting that taxidermy was necessary only as a method to preserve specimens from decay. Swainson argued that the method of preparing a skin for study was closer to nature than the artificial process of preparing a mount.[32] But there is arguably nothing more "natural" about the preparation of a study skin. In addition to the physical changes listed above, the Manchester Mandrill underwent a series of processes on entry to the museum to secure its status as a museum object. It was classified, categorized, labeled, and researched—assigned an accession number and coordinates that corresponded not only to its species and provenance but also to the collection world in which it had been placed. This process of signification did not end with the object's accession. Despite the initial efforts to render it stable, over the next one hundred years the Manchester Mandrill would be subject to a host of further changes to its artifactual, material, and scientific status within the museum (fig. 2).

## MUSEUM AFTERLIFE

When the Manchester Mandrill entered the museum at the beginning of 1909, it entered a thriving department.[33] Until the late 1920s, zoology was the driving intellectual discipline within the museum, reflected in staffing structure, gallery displays, and research output. In the precise year of the mandrill's accession, the museum's commitment to the research function of the zoology collection was demonstrated by the creation of a new space in the southwest corner of the museum "for the use of Students engaged in the study of the Zoological Specimens."[34] The Manchester Mandrill is highly likely to have occupied this space after its accession—its immediate environment one of active research and inquiry. But by the late 1920s, museums and their collections were no longer the principal sites for the production of knowledge in the life sciences. Further into the twentieth century, the university's Zoology Department began to distance itself academically from the museum, and study skins such as the Manchester Mandrill no longer played a key role as tools of research. As far as the recollections of past and present staff extend, the Manchester Mandrill has spent at least the last fifty years in storage. But rather than remaining static or lost, this period of apparent inactivity contains

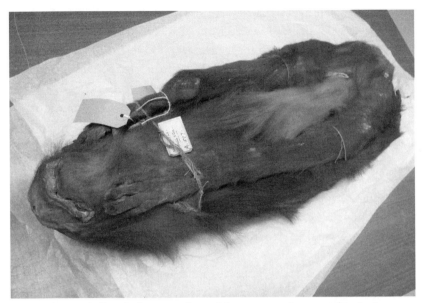

FIG. 2. The mandrill study skin, Manchester Museum accession A.597. (Author's photograph; copyright the Manchester Museum, University of Manchester)

its own dynamic, suggestive of the changing status of the zoological collection at Manchester.

Roy Garner, zoological conservator at the museum for more than thirty-five years before retiring in 2008, recalls the location of the mandrill in the early 1970s, when it was stored among a stack of miscellaneous skins behind 8-feet-high (2.5-meter) barriers closing off the current sea cow and Australian mammal section of the Mammal Gallery. These barriers, and what they were hiding, had been in place since the early 1960s. After that, the skin was kept in a drawer beneath a display case in the old Mammal Gallery. The next storage space was "a stereotypical 'museum store'—dark, dirty, with leaky steam pipes running through it." Following this, the skin was in temporary storage in the old Botany Gallery and then in storage adjacent to the old Conservation Department. In the late 1990s, the museum underwent major redevelopment, and all the collections were moved to temporary storage away from the museum for several years.[35]

The mandrill skin carries the material evidence of this journey through storage as particular environments left their imprint. The most significant of these is a small number of larvae cases—the remnants of an infestation that may have occurred up to sixty years ago. Fortunately, the skin's conditions were never bad enough for long enough to cause major deterioration, and considering the variety of conditions to which it has been exposed in the last one hundred years, the mandrill has remained in remarkably good condition. Today, the skin sits on the lowest level of a roller-racked shelving system in the basement mammal storage area. The environment is temperature- and light-controlled, in line with the best collections care that the museum can afford. In contrast to the temporary and inadequate storage of earlier years, these conditions demonstrate the recognition of preventative conservation in sustaining the stability of museum objects.

According to the testimonies of past and present museum staff, it is unlikely that the mandrill skin has been used as a tool of "study" for at least the last fifty years. While this owes partly to the general decline in object-based inquiry over the twentieth century, research and collection at Manchester has tended to focus on the bird skins, which represent a greater taxonomic range.

The "dis-use" of the Manchester Mandrill differs with the museum afterlives of mandrill specimens housed in the Natural History Museum (NHM) in London. The NHM holds twenty-six mandrill specimens, including seven study skins. Richard Sabin, senior curator of the Mammal Group (and fellow contributor to this volume), emphasizes the extent to which the NHM collection remains a vibrantly active site for research. The mammal collection alone currently averages around eight to ten research-related visits a day. Recent advances in specimen-based inquiry include the extraction of DNA from hair and skin samples and the analysis of stable isotopes from within bones to reveal previously hidden information about an animal's diet or habitat. Primates are one of the most consistently researched areas of the NHM's collection, and the mandrill specimens are regularly accessed.[36] The contrast in the "use" of the Manchester Mandrill skin and the specimens in the NHM is stark, though unsurprising given the relative status of each institution. Although it has faced its own uncertainties, the NHM has maintained its position as the leading center for specimen-based research in the difficult decades that followed the 1950s by steadily increasing its research output. Though the public galleries still perform an essential and successful function, the vast majority of work now occurs behind the scenes. The new Darwin Centre, designed to expose the research work of the museum to the public, suggests that research is now the true "dinosaur" attraction at the NHM. Like many successful regional museums, the Manchester Museum has survived by turning to its public—focusing resources on education, social inclusion, and its program of exhibitions. While not center stage, research is still a core component of the natural collections at Manchester. Although the mandrill skin has failed to attract the attention of researchers in the last half century, its status as an object of science may well be revived by advances in specimen-based inquiry such as those carried out on the collection at the NHM.

While the Manchester Mandrill skin has spent the last one hundred years in a succession of museum stores, an alternative life was lived by its mounted counterpart on the gallery floor. Until late 2010, a male mandrill grimaced aggressively from behind the glass of his case in the Animal Life Gallery of the Manchester Museum. The mandrill shared this space with a selection of primates, and like the many mounts on the gallery floor, he was tasked with representing an entire species. The majority of cases in the gallery followed this mode of classification and display, recalling the taxonomic methods of the nineteenth-century museum. But while this museological time capsule remained relatively untouched, public opinion regarding the representation of animals shifted around it. Since the 1980s, museum taxidermy has become increasingly problematized. As Patchett and Foster have argued, the stuffed and mounted skins of animals are now contested sites, viewed variously as "historical curios," "obsolete relics," or "uncomfortable reminders of past scientific and colonial practices that sought to capture, order and control animated life."[37] But to what extent do these shifts in public and academic opinion impact upon the many museum objects that remain out of sight? Are the study specimens that comprise the majority of our natural history collections any less open to reinterpretation because of their scientific credentials?

The experience of researching the Manchester Mandrill has suggested that study skins carry the potential to be equally, if not more, provocative. When I first asked to see the mandrill skin, Manchester Museum curator Rebecca Machin brought it to the museum's resource center on a trolley shrouded with tissue paper. Rebecca was anxious that museum visitors and school groups should not be exposed to a potentially disturbing sight. The animal's stuffed eye sockets, crudely sewn mouth, and folded legs present a striking contrast to the mandrill mount on the gallery floor. The gallery mount confronts the public with a ferocious snarl—the blue and red of its skin coloration vivid with life. Like the heroic Alfred in the Bristol Museum (see Hannah Paddon's essay in this volume), the mandrill meets your gaze, inviting you to imagine that it is still a living, breathing animal.

Untapped of its potential as a tool of scientific research, the study skin speaks principally of death. Like the hen harrier whose "biogeography" has been charted by Patchett, Foster, and Lorimer in their essay in this volume, the hollow eyes of the Manchester Mandrill stubbornly refuse to look back. One has to work harder to imagine the former lives of these creatures that somersaulted through the skies, prowled the rainforests of West Africa, or occupied the cages of a provincial zoo. But the skins have an arresting charisma of their own. The desiccated bodies of the mandrill and his fellow primates that line the shelves of the mammal storage area are disturbingly similar to the mummified human remains on display elsewhere in the museum. Like the mummies, they are absolute in their deadness. Their particular likeness to humans further forces us to question the way we represent death in the back rooms and front rooms of the museum. For Rebecca Machin, the methods of preservation and presentation of many of the mammal skins indicate a lack of respect. Machin reflects that this may be reason in itself to display them alongside their lifelike counterparts on the gallery floor: "Taxidermy mounts allow you to believe that the animal is still alive and well, whereas you cannot avoid thinking about our relationship with other species when you look at a folded-up mandrill."[38]

It is with Machin's thought that the mandrill's journey to date is concluded. The purpose of this biographical study has been to explore the potential of a single zoological specimen to unlock a host of cultural, institutional, and disciplinary narratives. The aim throughout has not been so much to give life to a particular animal, but to use the trajectory of the Manchester Mandrill to unravel the entwined histories of nineteenth-century animal representation, the trade in exotic animals, zoological gardens, and museum practice past and present. Throughout the mandrill's afterlife, its status has been determined by its relationship to the people who have worked, visited, or failed to visit the museum. While other objects were singled out for repair, research, or display, the mandrill languished in a succession of museum storage spaces, seemingly inert and lacking in human attention. But far from stagnating, the skin's movement, in and out of drawers, behind barriers, and on- and off-site contains its own activity. It is a biography that intersects with the institutional biography of the Manchester Museum and the people who have encountered the skin during the last one hundred years. When the mandrill reentered the museum from outside storage in the late 1990s, it entered a very different institution from that of 1909. Outside the museum, the mandrill's meaning remains in flux as we continue to question our problematic relationship with the natural world. Even for the curatorial staff of the Zoology Department today, the sight of the preserved skin is an uncomfortable reminder of a human-animal relationship to which they no longer subscribe.

NOTES

This essay shares a title with an article by Helen Rees Leahy on the display of human remains in museums: "Under the Skin," *Museum Practice* 43 (2008): 36–40.

1. William Smith, *A New Voyage to Guinea* (London: Nourse, 1744), 51.

2. Carl von Linné, *Systema naturae per regna tria naturae,* 10th ed., 2 vols. (Holmiae: Salvii, 1758–59).

3. Georges Cuvier, *The Animal Kingdom: Arranged in Conformity with Its Organisation,* trans. H. M'Murtrie (1817; New York: Carvill, 1833), 57.

4. William Bingley, *Animal Biography or Popular Zoology* (London: Rivington, 1829), 73.

5. Harriet Ritvo, *The Animal Estate: The English and Other Creatures in the Victorian Age* (Cambridge: Harvard University Press, 1987).

6. J. G. Wood, *The Illustrated Natural History: Mammalia* (London: Routledge, 1865), 76.

7. John Hogg, *The Parlour Menagerie* (London: Hogg, 1875), 147.

8. This story appeared as an anecdote in *Diary or Woodfall's Register,* 13 August 1790; *World,* 6 May 1791; and *Public Advertiser,* 10 May 1791.

9. For discussions on the wider cultural resonance of the drowned "fallen woman," see Deborah Epstein Nord, *Walking the Victorian Streets: Women, Representation and the City* (Ithaca: Cornell University Press, 1995); and Lynda Nead, "The Magdalen in Modern Times: The Mythology of the Fallen Woman in Pre-Raphaelite Painting," *Oxford Art Journal* 7 (1984): 26–37.

10. Wood, *Illustrated Natural History,* 76.

11. Ibid., 74.

12. The menagerie at Exeter Exchange on the Strand in London was a popular attraction from the 1770s (see Richard D. Altick, *The Shows of London: A Panoramic History of Exhibitions* [Cambridge: Harvard University Press, 1978]).

13. Accounts of "Happy Jerry" appear in James Rennie, *The Natural History of Monkeys, Opossums and Lemurs* (London: Knight, 1838); Thomas Brown, *Biographical Sketches and Authentic Anecdotes of Quadrupeds* (London: Fullarton, 1831); and

William Jardine, *The Naturalist's Library* (Edinburgh: Lizars, 1833). Jerry was do-nated to the British Museum in 1855, and his skin remains in the collection of the Natural History Museum, catalogue number ZD.25a.

14. To trace the mandrill's route from Africa to Belle Vue, it was necessary to give the animal a likely age at the point of death. Dr. Joanna Setchell, a primatologist at Durham University, used measurements and photographs to estimate an age of six or seven years old (Joanna Setchell, e-mail message to the author, 28 July 2009).

15. Belle Vue ledger, F.2.2: Animal lists inc. acquisitions, purchases and losses, Belle Vue archive, Chetham's Library. (Note that these records date only until 1907.)

16. *Era,* 29 June 1889.

17. The remaining mandrill was purchased from Mrs. J. Smith, Rochdale, in 1903. Like many of the Belle Vue monkeys, this animal is likely to have been kept as a pet before being sold to the zoo (Chetham's Library Belle Vue Collection, ledger F.2.2).

18. Ibid.

19. Cited in Robert Nicholls, *The Belle Vue Story* (Manchester: Richardson, 1992), 8–10

20. *Guide to the Belle Vue Zoological Gardens near Manchester with a Description of the Colossal Picture* (Manchester: Heywood, 1872), 9.

21. *Guide to the Zoological Gardens, Belle Vue, Manchester* (Manchester: Belle Vue, 1882), 6.

22. *Official Guide to the Zoological Gardens, Belle Vue, Manchester* (Manchester: Belle Vue, 1893), 10.

23. *Official Guide to the Zoological Gardens, Belle Vue, Manchester* (Manchester: Belle Vue, 1895), 10.

24. "Notes of Animals, Conditions, Habits etc.," 1908–14, Chetham's Library Belle Vue Collection, ledger F.5.4.

25. Chetham's Library Belle Vue Collection, photograph F.4.4.1.xiii no. iii.

26. Belle Vue–Manchester Museum correspondence, Manchester Museum Zool-ogy Archive ZAC/1/85/7; ZAC/1/85/10; ZAC/1/85/12; ZAC/1/85/17; ZAC/1/85/18.

27. "Notes of Animals, Conditions, Habits etc.," 29 December 1908.

28. James Jennison to William Evans Hoyle, 1 December 1908, Manchester Mu-seum Zoology Archive ZAC/1/85/9.

29. Mammal Accession Register, 1889–2009, Manchester Museum Zoology; *Manchester Museum Report 1908–1909* (Manchester: Sherratt and Hughes, 1909), appendix I, sec. 2.

30. Hardy was macerating skeletons and preparing birds and insects in the late nineteenth and early twentieth centuries, and as he later prepared the skin and skel-eton of a rhinoceros, it is likely that his skills may have extended to the preparation of a mandrill skin in 1909 (see Manchester Museum Committee Minutes vol. 1, 1 October 1890 and 24 March 1893, Manchester Museum Central Archive; *Manchester Museum Report 1900–1901* [Manchester: Cornish, 1901], 4; *Manchester Museum*

*Report 1902–1903* [Manchester: Cornish, 1903], 9; and *Manchester Museum Report 1917–1918* [Manchester: Manchester University Press, 1918], 5).

31. To extract the processes that Hardy or his colleagues used to transform the mandrill from dead animal to museum object, the skin was examined in 2009 by Roy Garner, zoological conservator at the Manchester Museum, 1972–2008. Garner recalls that as recently as the early 1970s there was a box in the Zoology Department labeled "Taxidermy Mixtures." This contained jars of "Taxidermy Soap," "Arsenical Soap," bars of old-fashioned laundry soap, and "White Arsenic" powder. The latter had a supplier's label which dated from the early 1900s (Roy Garner, e-mail message to the author, 29 June 2009; Roy Garner, interview by the author, 6 August 2009, audio recording).

32. Cited in Karen Wonders, *Habitat Dioramas: Illusions of Wilderness in Museums of Natural History* (Stockholm: Almqvist and Wiksell, 1993).

33. Samuel J. M. M. Alberti, *Nature and Culture: Objects, Disciplines and the Manchester Museum* (Manchester: Manchester University Press, 2009).

34. *Manchester Museum Report 1908–9* (Manchester: Sherratt and Hughes, 1909), 5.

35. Roy Garner, e-mail messages to the author, 25 June, 30 June, 1 July; Roy Garner, interview by the author, 6 August 2009, audio recording.

36. Richard Sabin, senior curator, Mammal Group at the Natural History Museum, interview by the author, 27 July 2009.

37. Merle Patchett and Kate Foster, "Repair Work: Surfacing the Geographies of Dead Animals," *Museum and Society* 6 (2008): 98.

38. Rebecca Machin, curatorial assistant (Natural Environments) at the Manchester Museum, e-mail message to the author, 28 July 2009; Rebecca Smith [Machin], "Ethical Implications of the Display of Non-Human Animal Remains in Museums" (master's thesis, University of Manchester, 2007).

RACHEL POLIQUIN

# Balto the Dog

alto (ca. 1914–1933) was a black Siberian husky and the lead sled dog on the final leg of a desperate journey in the winter of 1925 to carry the diphtheria antitoxin into the icebound town of Nome, Alaska. The extraordinary 674–mile (1,085–kilometer) run—through blizzards, across a frozen inlet, and in temperatures that dipped below minus sixty degrees—kept the nation enthralled for five-and-a-half days. Children were dying in Nome, and it would be the heroism of men and dogs that would save them. Balto was one of more than 150 dogs, and his musher, Gunnar Kaasen, was one of twenty men who relayed the serum into Nome. Another far more celebrated driver and his legendary lead dog, Togo, were perhaps the real stars of the rescue run, but by a strange twist of fate, it was Balto and Kaasen who delivered the antitoxin to Nome and garnered all the fame. Balto became a media darling and a national hero. He starred in a short movie. He posed with the actress Mary Pickford. He has a bronze statue in New York's Central Park (fig. 1). He is the hero of more than one children's book, including Natalie Standiford's *The Bravest Dog Ever: The True Story of Balto.*[1] And sixty-two years after his death, he had the leading role in the 1995 Universal Pictures animated movie *Balto,* which creatively retells the story of the "Great Race of Mercy," as the Nome serum run has become known.

Like many heroes and celebrities, Balto was honored in life because he ex-uded qualities that are admired in all animals—human or otherwise. Balto was strong, brave, determined, and an incredibly capable athlete. In the 1920s,

FIG. 1. The statue of Balto, 1925, by Frederick Roth, in New York's Central Park. (Copyright Uris; freely licensed via Wikimedia Commons)

no machine had yet been invented which could compete with the speed, agility, environmental awareness, and endurance of a well-trained dog team. Sled dogs have never been mere tools of winter locomotion but are partners and frequently leaders of the sled. A good lead dog with his near sixth sense for dangerously cracked ice or hidden crevasses is often the only reason a human musher and the entire team survive. A dog's incredible sense of smell can pick up a trail several feet below new snow in whiteout conditions. Certainly Balto saved Kaasen more than once on their race to Nome. When snowdrifts forced the team off the regular route, Balto found an unfamiliar path in the pitch-darkness of a blizzard by scent and feel alone. And several hours later, when the blowing snow was so thick that Kaasen could barely even see his dogs, Balto took complete control and navigated the team across the blinding landscape.[2] As the *New York Times* reported on 3 February 1925, Kaasen gave all credit to Balto: "He said the last leg of the relay would have been unsuccessful if Balto had not been on the team."[3] After all, it is Balto, not Kaasen, who is memorialized in bronze.

In death, Balto continued to be venerated. But in addition to the movies, the monuments (there are at least three Balto statues), and the press coverage, Balto was honored by a memorial which is never—or only exceptionally rarely—considered for human heroes. After his death in 1933, Balto was skinned, stuffed, and displayed at the Cleveland Museum of Natural History, where he remains today, 3,500 miles (5,600 kilometers) from his hometown of Nome. Although the sun has bleached Balto's preserved hide from its original jet-black to a subdued mahogany brown, he still exudes all the qualities of a good sled dog: alert, sensitive, willing, and ready.

On the one hand, Balto was preserved because he was more than "just" an animal. He was a courageous and intelligent savior, accomplishing what no human could have done alone. In extreme winter conditions, a hard divide between master and beast does not exist for sled teams. Balto was a partner with sensitivities and skills far beyond human capabilities. The mission itself is inconceivable with any other species. Bulls or rabbits, gorillas or cats would hardly have been able to partner physically and emotionally with humans to accomplish such a heroic mission. And perhaps dogs are always more than "just" animals. They are our companion species, our ancient partners in work and life, and probably humans' first nonhuman friends.

On the other hand, Balto was preserved because he was "just" an animal. No one would have ever suggested that Kaasen be similarly memorialized. Human heroes are not taxidermied. They are not skinned. Their skins are not refashioned in alert and lively postures. That dubious honor is reserved exclusively for animals.

Certainly there are examples of famous humans who have been preserved after death. Chairman Mao Zedong and Vladimir Lenin are both embalmed and on public display. When Argentine First Lady Eva Perón died, there was a push to display her embalmed body in a monument to the Argentinean worker. But there are several differences between such embalmed humans and taxidermied animals. To state the obvious, exceedingly few humans are embalmed for long-term display: a human has to be a saint or truly notorious to be considered worthy. Second, embalmed bodies are by and large whole, fleshy bodies, or at least more whole and fleshy than taxidermied skins. And, strangely or not, humans are almost always posed lying down with their eyes closed and their hands demurely folded or lying softly at their sides. It is the pose of a twilight death and engenders that instinctive hushed respect reserved for the sleeping and departed. In contrast, animals are almost always positioned in the action of life, with sparkling glass eyes to enhance the realism of the refashioned liveliness and—crucially—to camouflage the death. With animals, artistically crafted liveliness is on display. With humans, immortal sleep.[4]

But so in life, as in death: dogs are never really "just" animals. Despite the menagerie of taxidermy on display at the Cleveland Museum of Natural History, Balto stands out as something different and not merely because he is an animal celebrity. In fact, any taxidermied dog in any context will stand out as something different.

PRESERVED DOGS

Part of the reason that dogs are not regularly on display in natural history museums is that no one breed could offer a typical representation of the species. From Great Danes and dachshunds and teacup poodles to malamutes, dogs are too astonishingly variable to fit into the regular exhibition constraints of most museums. In the very rare cases in which dogs are displayed, more than one is almost necessarily required.

One such rare example is the Dog Collection on display at the Natural History Museum in Tring, which includes an impressive eighty-eight specimens. Almost all of the dogs were champions of their breeds and kennel club winners. Several were celebrated racers such as the greyhounds Mick the Miller and Fullerton, both of whom won endless races and the hearts of British racegoers. Most of the dogs date to the beginning of the twentieth century and were collected by the zoologist Richard Lydekker in order to create historical documentation of the ever-expanding number of breeds and to offer the public a visual education in selective breeding. The Dog Collection is displayed in a long, glass gallery case which allows almost all the dogs to be seen in a single—and rather singular—look.

The view is not the same as that offered by a gallery of ungulates or birds of prey. What the Dog Collection offers is a remarkable vision of humans' ability to fashion and refine canine (and by extension, the broader kingdom of nature's) shape, size, color, and aptitude by selective breeding.[5] Since all dogs are the same species, no dog is unmarked by human desires and ingenuity. Perhaps more than any other species, dogs are intimately and forever entwined with human culture.

From a more intimate perspective, the Dog Collection's dead dogs will always be more emotionally provocative than a gallery of deer or eagles. The lively display of a dead dog is somehow almost unseemly. A dog is never "just" an animal, a species, or a specimen. Even an unknown dog hovers on the border of friendship and intimate companion. Interestingly, most of the dogs are posed in the classic posture of a natural history specimen—standing and attentive but with a neutral gaze. Several of the dogs are lying down or sitting, and a few are disconcertingly mounted "trophy style" with their heads on a plaque. Why is this so jarring? Perhaps the human-dog bond is too intimate for such postmortem bodily invasion. Or perhaps the unsettling aura of a stuffed dog arises because most viewers know the manners and movements of dogs far more intimately than the manners and movements of zebras, buffalos, and pelicans. The uncanniness of all taxidermy—that disquieting liveliness within death—is necessarily heightened with dogs and other close domestic companions. We know exactly how they would shake themselves awake from their immortal sleep.

In an exhibition on the cultural history of taxidermy I curated for the Museum of Vancouver, Ravishing Beasts: The Strangely Alluring World of Taxidermy, a deceased pet dog named Lucky was displayed in an introductory section describing the eight genres of taxidermy.[6] The section was designed to provoke a variety of reactions from visitors (disgust, sadness, humor, neutrality) in order to emphasize taxidermy as a highly nuanced cultural practice. Hunting trophies, preserved extinct species, and stuffed pets are not the same sort of objects, and each reveals diverse (dis)connections with the natural world and arouses vastly different responses. The genres included a hunting trophy (a leopard head mounted by the great Victorian taxidermist Roland Ward), a natural history specimen (a pygmy owl on a stump), a wonder of nature (an albino skunk), an item of home decor (an elephant-foot table), a fraudulent animal (a jackalope), an extinct species (a huia bird), a scene of anthropomorphic taxidermy loaned from the Torrington Gopher Hole Museum (depicted a gopher at a train station carrying a tiny gopher-sized suitcase), and a stuffed pet (Lucky). From a cautionary tale against the misuse of nature to darkly humorous transformations, from pets to trophy kills, the significance of any piece of taxidermy is framed by particular social, aesthetic, scientific, and ideological concerns.

Lucky was among the most talked-about and most provocative creatures on display. He entranced visitors. He horrified visitors. His glass case was perpetually smeared with children's hand- and nose prints. One hundred and twenty animals were exhibited: an enormous moose, a white rhino head, hummingbirds, Kermode bears, a platypus. But it was Lucky who roused the most questions. Who was he? How did he die? When was he stuffed? In no way did Lucky blur together with the natural history specimens or the hunting trophies on display. Lucky stood out because, as a close domestic companion, he raises queasy emotional and ethical uncertainties about the practice of taxidermy in general.

A taxidermied dog stands apart from other species in one more sense. Without exception, dogs are taxidermied because they were somehow exceptional in life. They were champions of their breed, extraordinary athletes, heroes (whether voluntary or not), film stars, or cherished family pets. A stuffed pet is perhaps more emotionally unsettling than Strelka and Belka, the canine survivors of Sputnik 5's orbital flight on display at Moscow's Memorial Museum of Astronautics or a preserved star such as Bullet the Wonder Dog. The German shepherd starred with Roy Rogers and his famous Palomino, Trigger, in Rogers's western television series. Both Trigger and Bullet were stuffed and displayed at the Roy Rogers and Dale Evans Museum along with a wide variety of memorabilia from Rogers's life and times.[7] But whether a dog is preserved for private sentiment or public remembrance, whether the motivating desire is personal loss or historical value, there is nothing random or indiscriminate in the decision. To see a stuffed dog is to know that the animal was somehow exceptional in form, behavior, and personality. Or, to put it another way, to see a stuffed dog is to see the underbelly of canine fame.

Balto is the most famous, but he is not the only dog from the serum run that was memorialized with taxidermy. Togo and his half-brother Fritz were also mounted after death as tributes to their heroism and effort. But is transforming an animal into a memorial of itself a proper tribute for an animal hero? That question is open for debate. However, preserving a hero's body certainly ripens into an unsavory homage if the animal is subsequently neglected. Togo and Fritz both barely survived their afterlife. Both were sadly abandoned to the back rooms of museum storage, unlabeled and unknown. It is only by strange twists of fate that the dogs were rediscovered, transported back to Alaska, and put back on display, Fritz at the Carrie M. McLain Memorial Museum in Nome, and Togo at the Iditarod Trail Headquarters Museum in

Wasilla, Alaska. As ever, true heroes often slip into darkened corners of our collective imagination, sometimes quite literally.

## THE SERUM RUN

In the 1920s, Nome was the northwesternmost city in North America and certainly among the most remote. The town had sprung up after a single gold nugget was found on the beach in 1898. Two decades later, no more gold had appeared, but the little town had persisted with about 1,400 residents and was the main commercial center for the ten thousand or so people scattered in small villages and mining camps across the Seward Peninsula.

Isolated on the far west of the peninsula on the edge of the Bering Sea, Nome was hardly an easy place to survive. For seven months a year, the town was an icebound outpost beleaguered by blizzards, ice fogs, and freezing temperatures. A few weeks after the last cargo ship of the year left Nome in October 1924, the Bering Sea was once again frozen into a block of ice, impassable by any ship, and the town was once again cut off from the world except for a telegraph line. There was, however, one route into Nome: a 674-mile (1,085-kilometer) dogsled trail that snaked from the town of Nenana in the middle of Alaska along the Yukon River and out around Norton Sound. During the winter months, Nome depended exclusively on dog teams to bring mail and other supplies out along the peninsula. The trail took about twenty-five days.

On the morning of 22 January 1925, few people outside of Alaska had ever heard of Nome. Within a few days, the tiny town was on the front page of national newspapers. An epidemic of diphtheria had broken out, and Curtis Welch, the only doctor for hundreds of miles, was stranded with only a small supply of outdated antitoxin serum. With five deaths, twenty confirmed cases, and at least fifty other locals at risk, time was not on the town's side. Curtis sent a desperate telegram to the United States Public Health Service in Washington alerting the nation of the outbreak and urgently requesting one million units of diphtheria antitoxin. Another bulletin was sent out for all points of Alaska, again requesting the lifesaving serum. As if by miracle, the Anchorage hospital had three hundred thousand units, enough to hold back a full-blown epidemic until more antitoxin could be shipped to Anchorage from Seattle. But how to get it to Nome? Sending the medicine by ship across the frozen Bering Sea was impossible, and the winter weather was too hazardous for airplanes. Dogsleds were the only route into Nome. A plan immediately solidified to transport the serum to Nome in far less time than the regular twenty-five-day mail run. The serum arrived in five-and-a-half days.

Nenana was three hundred miles north of Anchorage by train. Once the serum arrived in Nenana, a relay of dogsled teams would carry it to Nulato, halfway between Nome and Nenana. There, the legendary Norwegian dog driver Leonhard Seppala and his equally famous lead dog, Togo, would carry the package the remaining 315 miles (500 kilometers) to Nome. Nulato to Nome was the most dangerous stretch of the trail, particular if a driver took the shortcut across Norton Sound, notorious for its constantly shifting and breaking ice floes and frozen winds blowing in from the Bering Sea. The shortcut trimmed about 60 miles (100 kilometers) off the trail, but sudden cracks in the ice or blinding conditions had often sent dog teams to their death. Since Seppala was in Nome, he would cross the deadly Norton Sound twice: once to pick up the serum and again to carry it back home. But if one man could do it, it was Seppala. Nicknamed "King of the Trail," Seppala was a living legend. He had dominated the All Alaska Sweepstakes dogsled race. He was the fastest, boldest driver, and his dogs were the best in Alaska. In a harsh land where a lead dog can save a man's life, Togo had become something of a legend himself. Seppala and Togo would save Nome.

The story captured the American imagination. It had all the elements of a first-rate news drama: a life-and-death situation, a near impossible struggle against time, a hero in Seppala, perilous conditions, and the romance of the Far North. "Picture an Alaskan scene," the front page of the *Washington Post* read on the morning of 31 January:

A virulent epidemic breaks out . . . in the most isolated and remote community of that vast territory; and the medicament needed for coping with it can be obtained no nearer than many hundred miles away. There is no railroad nor other means of speedy communication; the trip being impracticable for an airplane. But human need cries aloud, and human valor answers. And so in an Arctic mid-winter a man sets out with a twenty-fold dog team and sledge, to traverse the frozen wilderness with the precious supply. With those same dogs he has won prizes in holiday racing competitions; now he pits their speed and stamina after the Pale Horse and his rider, league after league along the snowbound trails, amid perils and solitudes unspeakable, seeking to save the perishing.[8]

That man was Seppala. The Arctic hero was lauded as "a champion dog driver," "the greatest driver," the "undefeated sleigh musher of the North," the "champion musher of the North Country."⁹ All hearts were with Seppala and his dogs.

On 27 January at eleven o'clock, the first musher, "Wild Bill" Shannon, met the train in Nenana, strapped the twenty-pound package of serum on his sled, and headed west. The temperature had dipped below minus fifty degrees. By the time Shannon arrived at a midway roadhouse four hours later, it was minus sixty-two degrees. Parts of Shannon's face had turned black from severe frostbite, and three of his dogs would eventually die of exposure and exhaustion. But still the relay pressed forward. On 30 January, the *New York Times* ran an update that the serum was due to arrived in Ruby, 250 miles (400 kilometers) west of Nenana, by midnight. The next day, the *Washington Post* reported that the serum was on its way to Kaltag, a further 200 miles (320 kilometers) along.

And then, at the last minute, a change of plans. The epidemic had spread. No matter how fast Seppala's team could run, fresh recruits would be able to run the second half of the trail faster than a single team. Additional drivers were called and assembled at various roadhouses along the route. There was only one problem with the new plan: no message could be got to Seppala, who was already on the way to Nulato. In fact, Seppala—the Great Hope—almost missed the relay completely. But on the eastern edge of Norton Sound, Henry Ivanoff miraculously met up with the famed musher, despite brewing blizzard conditions. With a windchill temperature of minus seventy degrees, Seppala turned his dogs around and began crossing Norton Sound for the second time that day.

As Gay Salisbury and Laney Salisbury relate in *The Cruelest Miles: The Heroic Story of Dogs and Men in a Race against an Epidemic,* what happened next is one of those animal stories that is equally impossible to believe as to discredit. While crossing the sound, Togo suddenly reared and somersaulted back onto his teammates. Seppala went to investigate. No more than six feet ahead was open water. To his horror, Seppala realized the team was on an ice floe and was drifting out to sea. All hoped seemed lost. But several hours later, Seppala saw that the floe was shifting closer to shore. They drifted for nine more hours on a block of ice in a freezing ocean until the shoreline was visible, but it was still too far for the team to cross. In desperation, Seppala attached a towline to Togo's harness and threw the dog across the open channel, hoping that Togo, once on the other side, would be able to pull the floe to shore. Togo made it to the other side, but the line broke off his harness, and with that break, Seppala and his team were lost to the sea. And then Togo did the extraordinary. He jumped into the icy water, snapped up the broken line in his mouth, scrambled back on shore, and pulled the floe close enough for Seppala and his team to jump across to solid ground.

After a night's rest, Seppala carried the serum a further 60 miles (96 kilometers) to Golovin. Since picking the serum from Ivanoff, Seppala and his dogs had traveled 91 miles (146 kilometers), more than two-and-a-half times the average distance traveled by the other drivers. And then consider that Seppala had already travelled 170 miles (274 kilometers) from Nome before meeting Ivanoff. As Salisbury and Salisbury write: "This was done at top speed, in blizzard conditions over heaving ice. He and the dogs had survived the ruthless challenge of Norton Sound and saved at least a day off the critical time schedule."[10] Seppala's original leg—the Nome-to-Nulato round-trip of 630 miles (1,013 kilometers)—might have been cut in half, but Seppala had still accomplished something extraordinary. At Golovin, the package was handed to Charlie Olson. Olson handed it on to Gunnar Kaasen, who was to run the penultimate leg of the trail.

Seppala and Kaasen were both Norwegians, and both worked for the same mining company. They knew each other well, and Kaasen knew Seppala's dogs. In fact, Kaasen's team were all Seppala's dogs, including Balto. Seppala had bred Balto but always considered him to be second-rate. He was certainly not a lead dog. Kaasen had a different opinion.

Kaasen started out in a violent, bitter, blinding blizzard unaware that the relay had been called to a halt until the weather improved. He persevered through the whiteout and despite being blown into a snowdrift by icy winds and nearly losing the serum, Kaasen pushed on to Port Safety, where the last driver, Ed Rohn, was to carry the serum the final 20 miles (32 kilometers) into Nome. However, when Kaasen arrived at Port Safety, Rohn was asleep, presuming that Kaasen was waiting out the blizzard 40 miles away. Conditions had improved, and the team was running well. Rather than waiting for Rohn to prepare his dogs, Kaasen decided to carry the serum the final leg into Nome, arriving at five thirty in the morning of 2 February. Apparently Kaasen staggered off the sled and collapsed in front of Balto, muttering "Damn fine dog."[11]

The public response to the heroism of man and dog was overwhelming. Money poured in from across the nation to reward the teams for their momentous efforts. For a while, the news focused on Seppala. After all, his had been the stalwart heart on which the nation had depended. But, slowly, the fame turned. Despite the efforts of all twenty teams, despite the fact that Seppala and Togo had by far traveled the farthest and most treacherous distance, despite the fact that it was not supposed to be Kaasen who arrived in Nome, the team that carried the serum into the beleaguered town skyrocketed to fame. Perhaps the turning point was the reenacted photograph of Balto's arrival into Nome which was splattered across national papers. The other teams were not exactly forgotten, but it was Balto and Kaasen who were offered a movie deal by the famed Hollywood producer Sol Lesser. The thirty-minute movie was rather erroneously and provocatively titled *Balto's Race to Nome*.

Not surprisingly, Seppala was hardly pleased with the twist of fate. He was heartbroken that Togo had not received the recognition that he deserved and was particularly devastated that Balto, not Togo, was memorialized in Central Park (fig. 1). And admittedly, the very fact that New York raised a statue in the dog's—any dog's—honor is impressive. Of the twenty-nine statues in Central Park, Balto is the only memorial of a notable animal. Located on the main path leading north from Tisch Children's Zoo, the bronze sculpture is slightly larger than life. The inscription on the plaque reads: "Dedicated to the indomitable spirit of the sled dogs that relayed antitoxins 600 miles over rough ice, across treacherous waters, through Arctic blizzards from Nenana to the relief of stricken Nome in the Winter of 1925. ENDURANCE FIDELITY INTELLIGENCE." Somewhat improbably, a bronzed Balto posed ready in harness with tongue panting had become the face of Alaskan sled dogs.

The serum run is also now commemorated each March by the Iditarod Trail Sled Dog Race, which runs from Anchorage to Nome over 1,150 miles (1,850 kilometers) of the some of the most treacherous terrain in the world. The Iditarod first ran to Nome in 1973, after two short races on parts of the trail in 1967 and 1969. Admittedly, the Iditarod was conceived to commemorate the importance of all sled dogs in Alaskan history and to help save Alaska's vanishing mushing heritage. But the spirit of the Iditarod has always rested on the valor, courage, and endurance of Balto, Togo, Fritz, and the other dogs of the great Race of Mercy in the winter of 1925.

CANINE SOUVENIRS

Despite a few hard years, Balto's fame continued unabated through his life. After the unveiling of the statue, Kaasen sold Balto and the team to a vaudeville show in Los Angeles, where they lingered in certain degree of misery and anonymity. But Balto was not fated to be neglected. While visiting the city in 1927, George Kimble, a former prizefighter turned businessman, was shocked to discover Balto and six of his teammates unhealthy, underfed, and suffering in the sideshow act. He decided to save Balto and his teammates by bringing them to his hometown of Cleveland. The owner agreed to sell the dogs for two thousand dollars, but only on the condition that the money could be raised in two weeks.

The price was steep and time was short, but Kimble was determined. Enlisting the support of local radio stations, the Animal Protective League, and the Western Reserve Kennel Club, Kimble held a fund-raising drive. Within ten days, the Balto Fund easily collected more than $2,300 from the good citizens of Cleveland, including the children. Many decades later, Emerson Batdorff recalled the outpouring of animal compassion he felt as a seven-year-old: "The teacher put it very simply so that everyone could understand: 'These heroic dogs, the dogs that went through the terrible snowstorm when no one else could get the medicine to Nome and save the people from diphtheria, are hungry and have no one to pet them.' It certainly convinced me—and everyone else in the room. That is how I happened to help bring Balto and the other sled dogs to Cleveland. As I remember, I gave 5 cents."[12] On 19 March 1927, Balto and his surviving teammates—Fox, Sye, Billy, Tillie, Moctoc, and Alaska Slim—pulled a sled on wheels down Cleveland's main street, lined with a jubilant crowd who still remembered the dogs' heroism.

Balto lived out the remaining six years of his life in relative ease at the Brookside Zoo, now the Cleveland Metroparks Zoo. In the spring of 1933, Balto was in extremely poor health: deaf, blind, and crippled by canine arthritis. Balto was put down by a kindly local veterinarian (Dr. R. R. Powell, a member of the Cleveland, Ohio, Balto Committee). He died at 2:15 PM on 14 March 1933. After his death, he was carefully preserved by a staff taxidermist at the Cleveland Museum of Natural History and displayed with his original lead. Dead and stuffed but still loved, Balto had become a warm spot in Cleveland's civic history.

For the next sixty years, Balto spent a low-profile afterlife, mostly in cold storage to ensure his longevity with brief visits on public display every other year. In the late 1990s, however, Balto once became the center of national attention.

In January 1998, an Alaskan third-grader named Cody McGinn wrote a report on a book about Balto's heroic deeds in the 1925 diphtheria run. Most likely the book was Standiford's *The Bravest Dog Ever: The True Story of Balto,* which gave Balto the hero's role in the Nome serum run. Even though Togo had already returned to Alaska, McGinn was determined that Balto, "The Bravest Dog Ever," should return and live out his afterlife in his birth state. McGinn encouraged his second- and third-grade classmates at Butte Elementary School in Palmer, Alaska, to write letters to their local governmental representative, Scott Ogan, requesting help to bring Balto "home." Ogan was more than obliging and sponsored a resolution known as "HJR 62—Bring Balto Back to Alaska."

When the House Community and Regional Affairs met on 18 March 1998, the Butte schoolchildren had the opportunity to voice their opinions by teleconference. McGinn testified that Balto was "a hero to Alaska," and expressed his belief that Balto would be an excellent tourist attraction. Gia Homstad, an eight-year-old second-grade student, claimed that "even though Cleveland saved Balto from a side show which was not treating him very well," he should still be in Alaska. "Balto is a hero to all Alaskans," she continued, stating that "if Balto was able to choose between staying in Ohio or Alaska, he would choose Alaska because his story started here."[13] For his part, Ogan framed the resolution in terms of the use value of material history by arguing that

> if the preservation of the past is to have the most effect in helping us keep a full perspective on the significance of historical events[,] the preservation should be kept as close to the genesis of the event as possible. The question is, where is Balto most valuable and in what ways can his preservation best serve people[?] I believe the answer is by being here in Alaska where others may draw inspiration in the continuation of the celebrated trek to Nome called the Iditarod race, which he along with other members of his team have inspired. It is here the world turns its attention each year to commemorate Balto's feat. It is here the famous lead dog will have the most profound effect on the world and what the significance of the serum run still means.[14]

In other words, Ogan argued that Balto was not most importantly a famous dog but rather a monument to the hearty Alaskan spirit encapsulated by the Nome serum run and the annual Iditarod race. "The Last Great Race on Earth," the Iditarod is hailed as the "Spirit of Alaska," "a race extraordinaire, a race only possible in Alaska."[15] In a sense, the race captures the Alaskan identity as a place apart from the contiguous forty-eight states, and Balto literally embodied the race.

The Alaska legislature proceeded to pass the "Bring Back Balto" resolution. Cleveland refused to give Balto up. Their reasons were obviously of a different spirit. If Balto had saved Nome, Cleveland had saved Balto. The Cleveland Museum's director, James King, argued: "If we had just gotten Balto when he died and there was no association with Cleveland, I'd have no problem returning him. But because Balto spent more than 60 percent of his life in this city, the city helped rescue him, and he became a very big part of this city's history, he belongs here."[16] In other words, Balto made Cleveland citizens as proud of their involvement in Balto's recovery as Balto made Alaskans proud of their state's spirited history. King stood his ground but conceded to send Balto on a six-month visit to Alaska in October 1998, where the dog was displayed at the Iditarod Headquarters Museum.

SAD SOUVENIRS

But what about the other dogs? In 1926, Seppala took forty Siberian huskies including Togo and Fritz on a tour across America to race and generally to show off the dogs while the memory of the famous race was fresh in the popular imagination. Togo and Fritz remained and died in New England, suffering shades of obscurity in their afterlives before being rediscovered and brought home to Alaska.

Seppala retired Togo with a dog breeder in Maine. After his death in 1929, Togo was mounted for display in the Leon F. Whitney Dog Collection at the Yale Peabody Museum. Whitney was a local veterinarian, a dog breeder, and a student of genetics. He believed that a collection of champion kennel club dogs—the epitome of their breeds—would be an invaluable reference for future breeders. Known as the "Dog Hall of Fame," the collection was meant to create a physical record of canine evolution under selective breeding and to offer the general public a lesson in breeds of dogs. Togo was the only dog on display who was not a purebred champion. He was small for a sled dog, but he had the name and offered an exceptional image of the crossbreeding that had created the perfect sled dog. His father was half–Siberian husky, half–Alaskan malamute, and his mother was a Siberian husky imported from Siberia.

In the 1960s, the Whitney Dog Collection went into storage. But Togo, because he was Togo, eventually found a new home in a sled-dog diorama at the Museum of Shelburne in Vermont. There, however, Togo's afterlife took a turn for the worse. He was not protected by glass, and after two decades of incessant petting, he was reduced to a shabby hide. His tail and ears had lost all their hair. Unaware of the dog's history, the museum's director put Togo into storage and forgot about him. That would have been the end of Togo if Ed Blechner, a carpenter at the museum and an avid sled-dog breeder and racer, had not realized who the mangy creature had been in life. Inquiries were made in Alaska to see if anyone was interested in Togo. Everyone was interested. Pleas to rescue the dog poured in. After debating the various options, in February 1983 the Museum of Shelburne deaccessioned Togo and sent him home to Alaska. Because of the years of neglect, his tail and ears had to be replaced with those of another dog. Sadly, the patchwork is obvious.[17]

Fritz's circuitous route back to Alaska is an even stranger story. In 1929, Seppala sold Fritz to Beverly Sproul, a doctor in Maine. When Fritz died in 1932, Sproul had him stuffed. When Sproul died, a neighbor—Jacques Suzanne—acquired the dog and fabricated a story that Fritz had been part of Admiral Robert Peary's famous Arctic exploration. When Suzanne died, his possessions were auctioned off. Fritz, accompanied by Suzanne's erroneous narrative, was purchased for Frontier Town, one of the country's first theme parks located in upstate New York. When the park closed in 1998, Fritz was again put up for auction, this time amidst the jumble of amusement park wares under the anonymous listing of "Alaskan Husky Dog Mount in Glass Case." He was purchased by an antique dealer who had no idea of the dog's history. At that point, Fritz should have been lost. But someone was tracking him from a distance. In 2005, Natalie Norris and her brother bought Fritz without telling the dealer who the dog really was. A dog racer herself, Norris had followed Fritz's meandering and rather inglorious afterlife since he had been on display in Sproul's home. She had no interest in keeping the dog for herself but hoped to return him to Nome, almost eighty years after he had left.[18]

Balto, Togo, and Fritz were just three of 150 dogs that accomplished the impossible in the winter of 1925. We remember the three more clearly than the rest in part because they were memorialized with taxidermy. But, as ever, taxidermy is a dubious honor. It comes with no guarantee of future care. In contrast to Balto's easy course through death, the afterlife sagas of Togo and Fritz highlight several unfortunate features of taxidermied canine celebrities. The dogs were preserved as public memorials so that citizens might remember and appreciate what the dogs had accomplished. But unlike statues and biographies, a taxidermied animal requires almost constant care and attention. Unfortunately for Togo and Fritz, that responsibility became an unwanted burden. Plus, as with any souvenir from a deceased hero or celebrity—a piece of clothing, a worn writing desk—once the item becomes detached from its historical narrative, the cherished token quickly becomes commonplace. Separated from their heroic deeds, Togo and Fritz were consigned to oblivion (and perhaps to the queasier realm of a forgotten stuffed pet), and it was only by bizarre and improbable chance that either dog was rediscovered.

But a taxidermied canine hero is a very unique sort of souvenir. Through human craft, the animal is transformed and reimagined into a souvenir of itself while still being itself. The preserved bodies of the dogs are not representations of Balto, Togo, and Fritz: they *are* Balto, Togo, and Fritz (minus a tail, two ears, and all the inner workings). The dogs have been made to play a role, but the raw authenticity of their animal form continues to exert a visceral presence. This is true of all taxidermy: it is simultaneously a representation and a presentation of animal form. But again, a preserved dog is a very unique species of taxidermy. In contrast to almost all other preserved species, we know one thing for certain about a stuffed dog: it was not hunted. It either died naturally or was euthanized in old age. In either case, it lived, died, and was valued within a human world.

Any perceived divide between humans and other animals becomes exceedingly troublesome with dogs. Precisely because he was a dog, Balto captured the American imagination in a way that few humans ever could. Balto happily put his life in jeopardy for Kaasen. The raw affection of dogs, their enthusiasm, superhuman capacity, and full commitment to the moment set canines apart from almost every other species, including most humans. It is because we love dogs, and they love us, that they are sometimes given an afterlife with taxidermy. And yet, it is also because we love dogs that any such postmortem preservation can seem tasteless and unsettling.

The story of how Balto, Fritz, and Togo became famous is an extraordinary story of life, death, and heroism. It is a story of how any dividing line between humans and other animals can evaporate or harden depending on the circumstances. Ultimately, it is a story about the fickleness of fame, the strange ways we remember our animal heroes, and what happens when we forget.

NOTES

1. Natalie Standiford, *The Bravest Dog Ever: The True Story of Balto* (New York: Random House, 1989).

2. Gay Salisbury and Laney Salisbury, *The Cruelest Miles: The Heroic Story of Dogs and Men in a Race against an Epidemic* (New York: Norton, 2003), 222–24.

3. "Serum Arrives Frozen Solid," *New York Times,* 3 February 1925.

4. Of course, there are exceptions to this rule. Jeremy Bentham is perhaps the most notorious. He requested in his will that his body be preserved as an "Auto-Icon." According to his wishes, his body was preserved seated in a chair and dressed in the clothes Bentham wore in life. The preservation was so well executed that his friend Lord Brougham claimed that the likeness was "so perfect that it seems as if alive." His head was later replaced with a wax replica, although his embalmed head is kept safe in a box nearby. He is still on display in the main building of University College London. For more on Bentham's afterlife, see C. F. A. Marmoy, "The 'Auto-Icon' of Jeremy Bentham at University College, London," *Medical History* 2, no. 2 (1958): 77–86.

5. Kim Dennis-Bryan and Juliet Clutton-Brock, *Dogs of the Last Hundred Years at the British Museum (Natural History)* (London: British Museum [Natural History], 1988).

6. The exhibit Ravishing Beasts: The Strangely Alluring World of Taxidermy ran from 22 October 2009 to 28 February 2010.

7. After Rogers's death in 1998 and his wife's (Dale Evans) death in 2001, the museum was moved in 2003 to Branson, Missouri. However, Rogers and Evans's son has since closed the museum, and Trigger was sold at auction in July 2010 for a staggering $266,500 (Melissa Milgrom, "Triger Rides Again," *Salon,* 31 July 2010).

8. "An Arctic Epic," *Washington Post,* 31 January 1925.

9. "Seppala Making Record Dash," *New York Times,* 1 February 1925; "Dog Team Speeds over Frozen Sound with Aid from Nome," *Washington Post,* 1 February 1925; "Dog Team Carrying Serum to Nome Sought by Relays, *Washington Post,* 2 February 1925; "Final Dash Brings Antitoxin to Nome, But It Is Frozen," *New York Times,* 3 February 1925.

10. Salisbury and Salisbury, *The Cruelest Miles,* 214–15.

11. Ibid., 225.

12. Emerson Batdorff, "When Kids Saved Balto," *Cleveland Plain Dealer,* 25 August 1998.

13. The minutes of the meeting are posted online at www.legis.state.ak.us/basis/ get_single_minute.asp?session=20&beg_line=0199&end_line=0296&time=0802 &date=19980318&comm=CRA&house=H.

14. Ogan's comments are taken from his "Sponsor Statement for HJR 62," www .akrepublicans.org/pastlegs/spsthjr06203051998.htm.

15. The Alaskan spirit captured by the Iditarod is described in detail on the official website of the Iditarod at www.iditarod.com/learn/.

16. Brian E. Albrecht, "Cleveland to Stay Final Resting Place for Balto: Permanent Return of Heroic's Dog's Remains Is Unlikely Despite Alaskan Schoolchildren's Bid," *Cleveland Plain Dealer,* 14 February 1998.

17. See "Heroic Dog Is Home at Last," *Milwaukee Journal,* 3 March 1983. See also Earl J. Aversano's comprehensive history of the serum run, its dogs, and their afterlives at www.baltostruestory.com.

18. "Canine Who Helped Inspire Iditarod Is Back in Alaska," *Anchorage Daily News,* 24 February 2005.

MERLE PATCHETT, KATE FOSTER,
AND HAYDEN LORIMER

# The Biogeographies of a
# Hollow-Eyed Harrier

## JUST THE THING, ITSELF

Even in her reduced state—and before other words intrude—she remains a thing of the severest beauty (see fig. 1). Breast: a fine-weave swatch of caramel and crème. Wing feathers: close-plated, clean-edged, with arching white strips. Eyes: emptied, yet defined by a pale-colored patch, tapering to a hooked *V.* Primaries, when fanned as if for flight: ring-tailed with dark bars of sober brown, alternating with blocks of white. By their very nature, bird skins are featherlight; husky, and dry to the touch. Though long since ruffled, they still offer whispers of the airy life. For the pure love of aerobatics, these wings dipped, baffled, arrowed, buffeted, and flexed. Once, she threw caution to the wind.

The hen harrier's story is one of a fate sealed as soon as its name was conferred. Commonly known by countrymen of sixteenth-century shires as the "hen harrier," or "harroer," this was a bird nominated according to habits and tastes rather than looks.[1] The sworn enemy of free-ranging poultry, and by association their keepers, its ill repute for butchery overruled any aesthetic appeal. A penny bounty was placed on its head.[2] So it is that harriers must only have enjoyed true prelapsarian freedoms of the skies before man domesticated animals. Unloved for centuries. A breeding population harried back to westernmost isles, only recently recovering proper footholds on mainland

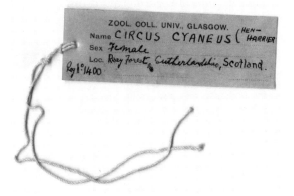

FIG. 1. Study skin of a female hen harrier, with museum label given as detail. This was used as a title image for Kate Foster's artwork *Disposition* (2003). (Image copyright Kate Foster/The Hunterian Museum and Art Gallery, University of Glasgow)

Scotland. And to this day, in spite of more widespread distribution: red-listed in status.

Given this troubled history, it is hard to offer a behavioral description of the hen harrier *apart* from humans. To begin to know just the thing itself, the briefest notes of introduction must suffice. The harrier is a ground-nesting bird. The female is renowned as "an exceptionally tight-sitter," and the doughtiest defender of her fledgling brood.[3] Nests, built in long heather and rank ground vegetation, are subject to predation by the red fox. Hunting in search of small mammals and birds, harriers favor open country, spying hillsides and working clear-felled ground. During spring come rituals of courtship, the male's tumbling and toying skydance, a spectacular signature in which he is sometimes joined by the female. During winter, outside the breeding season, they are known to congregate, using communal roosts.[4] These few things we know.

STUDY SKIN, STUDYING SKINS

Much attention has been paid to taxidermy mounts, their representation, individual histories, matter, and meaning.[5] Yet study skins, as hollowed-out scraps of lightly stuffed preserved skin, present less charismatic objects of study.[6] Garry Marvin has commented in response to taxidermy that due to their historical recording as scientific information, "each animal becomes a type, a token, rather than a unique individual."[7] This is especially true of a study specimen as unlike mimetic taxidermy reproductions, which often acquire unique afterlives in the museum setting, it is usually merely one of a mass of material exemplars for the distinct species that line the airtight drawers forming the empirical base of natural history study.[8] Anonymous, and for the most part kept in the dark, these could be thought of as largely forgotten and overlooked remnants of animal life.

Yet the aim of this essay is to return to one such remnant—the study skin of a hen harrier from the zoology collections of the Hunterian Museum, University of Glasgow—and make use of it. More specifically, the hen harrier study skin offers the departure point for a collaborative "biogeographical" study by an environmental artist (Kate Foster) and two geographers (Merle Patchett and Hayden Lorimer). While "biogeography" traditionally refers to a subdiscipline of geography concerned with mapping patterns of spatial distribution of species, we use it as malleable term which encompasses the different endeavors of artist and geographers interested in renewing the place of life in its multiplicity of human and nonhuman forms, processes, and connectivities.[9] More specifically for our joint work, the term titles a series of collaborative works and offers an alternative biographical practice.[10] Viewing zoological collections and specimens as *resources* for telling complex histories of human-animal encounter, we have, in both our individual and collaborative work, sought to emphasize the potential of working intimately with the unique histories of zoological specimens to elicit different kinds of knowledge and viewpoints about them beyond, yet still informed by their uses, in biological science. This particular study of a hen harrier study skin by the authors has evolved in response Kate Foster's original artwork *Disposition* (2003/2009) exploring the harrier specimen's unique history.[11] In this collaborative evocation, we aim to chart the hen harrier specimen's historical and contemporary "biogeographies" through life, death, and afterlife.

While Patchett undertakes the general narration of the present piece, this narration develops in correspondence with, and is interrupted by, Foster's artwork and ongoing research on the hen harrier specimen and species and upon Patchett's own theoretical understanding of what a specimen animal is and can do.[12] Lorimer's specific contribution takes the form of the opening creative text, yet through joint work with Foster and Patchett, we have collaboratively developed a "biogeographical" approach to the study and use of zoological collections and specimens, for which we make an argument in the conclusion.

In recent years, zoological museums and collections, with their vast quantities of specimens and samples in store, have often been cast, perhaps unfairly, as "conduits of disposal."[13] Yet the fact that these specimens are in storage attests to the fact that the museums do not want to get rid of them; they want to hold them for posterity—primarily because they are representatives of particular species but also because their possible further scientific usefulness has yet to become apparent. This kind of storage, for Kevin Hetherington, implies an "intention to hold something provisionally as absent" and that consequently disposal in this sense "is about the mobilisation, ordering, and arrangement of the *agency* of the absent." For example, even when an artifact, in this case a zoological specimen, is in storage and is not readily accessible, Hetherington argues it can still have effects through other media: "in the card index or catalogue those items may be just as visible as items on display."[14] Furthermore, in smaller collections, like the one at the Hunterian, their agency may be conducted through the curators themselves. For example, when Foster, who had already built up a relationship with the curator to use bird specimens for her artwork, was invited to do more work using the collections involving a contact of the curator's at the Royal Society for the Protection of Birds (RSPB), one specimen in particular suggested itself to the curator.[15] After some discussion, Foster and the RSPB contact decided to base the work around a hen harrier specimen as the species is an RSPB Scottish priority and the Hunterian had a number of their study skins in store. The curator remembered a harrier specimen that had additional data attached beyond the standard label information and brought it out on a tray with others for identification and closer study.

The hen harrier specimens were separately wrapped in a plastic bag, giving the impression of forensic specimens awaiting identification. And, as the RSPB contact made clear, this was a good analogy, as the species has been victim to a centuries-old wildlife crime. Classified as "vermin" by estate owners since the sixteenth century, they have been trapped, poisoned, and shot in their many thousands. Although supposedly gaining complete legislative protection by the 1954 Protection of Birds Act, they are still considered a threat to country-side economies which continue to privilege the pheasant and grouse over birds of prey.[16] The gamekeeper's secret cache, hen harriers (along with many other birds of prey) are now "dealt with" under the cover of darkness, then bagged up and burned, leaving no mortal remains.[17]

So to have a tray full of harrier study skins seemed miraculous consider-ing human action is usually aimed at extinguishing all physical trace. Thus, although only preserved skins loosely stuffed with tow, each one could be con-sidered to be a valuable "trace-bearing object" as "the thing so eloquently per-formed through the fragile presence of its hollow shell, its skin . . . is precisely that animal's reality."[18] Yet as Everest also makes apparent in this volume, it is precisely because of their simple preservation, their "hollowness," that cabinet skins, over their mimetically modeled display-taxidermy counterparts, more readily perform an animal's fatality and are thus, for some, far more "redolent of death."[19] Yet death, or rather the deadness of a study skin in this instance, is manifest both as a representational presence as well as a figural absence. This simultaneous closeness to, yet distance from, death connects back to Kevin Hetherington's account of disposal in that when things are held in a state of abeyance like in the case of a study skin, it prompts us to question "how we *ac-count* for or are held *accountable* by that which we have tried to dispose of but have left unfinished." Disposal in his account therefore becomes a question of the way in which we use or deal with the unfinished: "the aporial absent present . . . within our otherwise seamless representations." Thus rather than unquestioningly viewing the harrier specimens as representations of a particu-lar species of bird, their movement between categories of presence and absence disrupts such a reading, prompting us to question their remaining value. In this sense, they can be thought of as being "held at our disposal."[20]

Foster, since being first introduced to the collections at the Hunterian Zo-ology Museum, has been making use of the specimens placed at her disposal by the curator, describing her resulting artworks as "a series of interventions in the "after-lives" of zoological specimens."[21] These interventions are con-cerned with making tangential environmental histories that reflect, obliquely, on how natural history presents its subjects and on the contemporary value of these collections. Foster knew that, in the form of study skins, the harriers offered a starting point for investigating their individual histories as each was accompanied, as is standard practice, with their own "scientific label," a label which details, at the very least: scientific name, sex, and location where it was found or killed. Yet, as noted previously, the curator had remembered one specific hen harrier specimen that had an additional label attached, offering more for an artist to go on. The first label detailed scientific name, *Circus cyaneus;* the sex, female; and that the harrier was killed in Reay Forest, North West Sutherland.[22] Another indicated the specimen was acquired for the Hunterian Zoology collections in 1926 from Macpherson's Sporting Stores, Inverness. Together the labels therefore detailed the location where the harrier met its demise, the taxidermy workshop where it was "dressed," and the year in which it was added to the Glasgow Hunterian zoological collections.

The process of charting an artifact's biography is well rehearsed in scholarly research from various fields; however, as a working method it has, in the main, been employed to tell stories about the people who collected the artifact(s) or the institutions that encased them.[23] By tracing the harrier specimen's "biogeography," instead of repeating the anthropocentrically inflected and linear process of tracing a specimen's "biography" (which often connotes tracing the individual specimen's history within the museum), it becomes an act of tracing and placing it within the lifeworlds of human-animal encounter, cohabitation, and estrangement which brought it to and maintain its current situation as a representational scientific specimen. From this vantage point, "spaces of a life" are of interest, and personhood is recast as a constellation of events and encounters where animal and human lives lose neat beginnings and endings. Distinctions often drawn between "nature" and "culture" seem beside the point when the starting point of our investigations is that relationships between human and nonhuman worlds are inextricably entwined.[24] Such a viewpoint lets animals and artifacts gain agency, and human investigators are obliged to develop an eye for how we collectively shape, and are shaped by, nonhuman existences.[25]

In the remainder of the essay, we rework some contexts of the hen harrier specimen as "vexed dioramas"—using "field," "workshop," and "collection" as the pivots from which a series of excursions and returns will be made. The challenge of recovery necessitated that we purposefully assemble diffuse zoological and historical remains to tell of the harrier's biogeographies. By enacting a form of "salvage ethnography" in what follows, the harrier specimen, a taxidermist's workshop, and a museum collection provide us with various routes to investigate lives lived, things being done, and what is left behind.[26]

## FIELD: SITE OF DEATH

Reay Forest, the harrier's location of death (see fig. 2), was bought by the Duke of Westminster in 1921 from the Duke of Sutherland. The Duke of Sutherland had sold the land because it was no longer earning its keep from sheep farming, the activity for which the land was originally "cleared."[27] At a time when the Highlands were being systematically turned into a blue-blooded leisure park, the Duke of Westminster would have recognized the potential of turning the estate into a "deer forest"; not a forest at all but an area of enclosed moorland where deer and grouse could be "managed" and where southern "gentlefolk" would pay up to £100 a day for the privilege of shooting them.[28] The killing of birds of prey like a hen harrier and other named "predators" made up much of the everyday work of a sporting estate, and estate owners offered premiums for the destruction of such forms of "vermin." A thread of oral history, in the form of conversations with Willie Elliot, the son of the head keeper of the Westminster Estates in the 1920s and a retired salmon ghillie himself, enables us to offer a partial reconstruction of the events surrounding the harrier's death.[29] (While the term "ghillie" is recognized as patronizing nowadays, deriving from the Scottish Gaelic word *gile,* which translates to "lad, servant," it would have been used to describe the stalkers of the Reay Forest Estate at the time of the harrier's demise.)

According to Elliot, the bird that now belongs to the Hunterian collection would have been an "adventurer" from the Outer Hebrides who flew to the mainland looking for new territory. A welcome party of sorts would have met the harrier, as gamekeepers were under strict orders from estate owners to kill any potential game predators. Gamekeepers were even specially dressed to assist such assassination attempts. Estate tweeds were unique weaves designed to allow the disappearance of ghillies onto particular grounds. The story goes that ghillies were sent up on the hill in different tweeds for the estate owner to ascertain which made the best "camouflage cloth."[30] Hunter's of Brora, who manufactured Westminster Tweed (and many others), could therefore be thought of as a livery of camouflage for ghillies.

Elliot informed us that the Duke of Westminster kept a vault of tweed and would cut a "suit length" for each visiting guest, a custom only the wealthiest estates could afford.[31] As head ghillie, it was Elliot senior's job to take the duke's guests, freshly patterned, onto the hill between 12 August and 20 October, the estate's stalking season. Good stalkers were required not only to know the habits of deer intimately but also to know their "beat" like the back of his hand, and with Reay Forest being split into six beats, each covering twenty thousand acres, this was no easy task. Yet even if the stalker was well versed in both, there was no guarantee of a kill, and often ten hours on the hill would pass without a rifle being cocked. Spending long hours "spying" and tramping the heather together at close company and in changeable conditions, ghillie and guest knew for a certainty by the end of the day whether they liked each other or not.

It was during such a "beat" that Elliot reckons the hen harrier was shot. While the Reay Forest Estate offered only deer stalking and salmon fishing and thus had no grouse to protect, the killing of raptors and other birds of prey, which were classed as vermin, was enacted as an unquestioned part of good gamekeeping practice. The killing of "vermin," while viewed as gratuitous spoilage today, offered a way for ghillies and gamekeepers, whose wages were far from handsome in the 1920s, to make a lucrative supplementary income as premiums were offered by estate owners for their destruction.[32] Estate workers like Elliot's father, therefore, had their eyes well trained to recognize and shoot black-listed creatures as they could expect to receive up to an extra ten shillings in their pay packet for the killing of a raptor, the skin of which they could then sell to the taxidermist for twice as much again. This said, the killing of a hen harrier would have been a significant event as they were extremely rare in the area. It is for this reason that Elliot thinks his father did not dispatch the bird, as such a rare kill would have been a talking point. He surmises that a "gun" most likely killed the harrier—"gun" being the term used to describe a gentleman guest with a shooting lease for the estate. While traditionally the gun's most coveted quarry was the stag (with its impressive antlers for wall mounting), if it was a slow day on the hill, the killing of a rare bird like a harrier would have made for an interesting anecdote to be recorded in the gun's game book. Having bagged such a rare specimen, the gun would have sent the skin to Macpherson's Sporting Stores along with the rest of his haul so he could have a trophy made, and it is to there that we now move.

FIG. 2. Documentation of the artwork *Disposition* by Kate Foster (2003) in development: *a–b*—car journey from Glasgow to Sutherland; *c–h*—experimental location photographs on the Reay Forest Estate, Sutherland (not exhibited); *i*—site of the former Macpherson's Sporting Stores and taxidermy workshop, Inglis Street, Inverness. (Image copyright Kate Foster)

g

h

i

WORKSHOP: SITE OF TRANSFORMATION

The harrier was "dressed" by John MacDonald (1884–1969), a taxidermist at Macpherson's Sporting Stores, Inglis Street, Inverness. The store functioned as the supplier and taxidermist of choice to the Highland's huntin', shootin', and fishin' fraternity for more than eighty years, and thus played a central role in the town's social and economic life. Falling out of business in the 1970s, MacDonald's workaday material legacy was acquired by Inverness Museum, the only remainders of the most famous taxidermy workshop in Inverness, a town which had been a center for taxidermy for more than a century through its enormous trade in sporting mementoes.

The workshop contents, once on public display, are now stored. Removed from their original situation or documentary context, objects and materials can lose meaning and significance. Yet, by working "in the grain of things" and with a knowledge of taxidermy practice, the fragmentary objects and illegible materials once part of MacDonald's taxidermy workshop can be made to suggest "obsolete networks of use and affinity."[33] For example, an oversized needle, boxes of brown no. 24 artificial eyes, a saw with chipped teeth, and MacDonald's workbench covered in saw notches, strange oily stains, and patches of waxy buildup all reference the heavy labor involved in the main specialty of MacDonald's taxidermy work: the setting up of stag heads. A "Flit" fly sprayer and other Rentokil products are redolent of the fine line held between life and death in taxidermy practice, hinting at the pungent working environment that MacDonald would have had to endure. While fine paintbrushes, no. 4 black eyes, vivid paint colors, and Japanese wire gimp for "repairing bird's legs" all bring to mind the more delicate bird and fish taxidermy work in which MacDonald was also employed, it is the more personal items like MacDonald's spectacles and wonderfully appropriate deerskin apron that are most evocative. Caitlin DeSilvey, following Benjamin's theory of historical constellations, has pointed out that "potential awakenings" reside in objects and materials that people gather around them and eventually discard in the course of their lives, and that encounters with such discarded items can "propose empathetic connection with the people who made and handled them."[34]

A tape-recorded interview (now lost) with the firm's owner at its demise —Hamish Macpherson—also held at the museum, helps to further enliven the material remains.[35] Hamish had obviously known John MacDonald well, and in the interview, recorded around the time of the purchase of the workshop contents, he divulges stories and anecdotes about the man and his work. According to Hamish, MacDonald, who had worked at the firm well into his eighties, had "no equal for the range and quality of his work" in the town. Yet as stag heads were the prize most coveted by the constant stream of southern gentlemen traveling up on the twice-daily train from London during the shooting season, John MacDonald as the firm's veteran taxidermist had "mounted more deer heads than he or anyone else could remember." During an average September-October, the height of the shooting season, MacDonald had to deal with more than 350 items for mounting. This was made all the more demanding thanks to the firm's trade label promise: "Gentleman's private collections attended to on shortest notice." Hamish relays that such work could often be stomach churning as stag heads arrived somewhat "high" (even though partly preserved and wrapped in hessian) after the several days' journey it took to reach the store's workshop from moors to the North.

MacDonald's biggest single job came in the form of forty deer heads shot by the Maharajah of Alwar, who had the shooting lease of the Ceannaeroc and Ardverikie estates, which he wanted to be mounted and dispatched to India. And the maharajah was by no means the only high-status customer to engage MacDonald's services. The firm's customer lists, also kept at the museum, read as a "who's who" of the British aristocracy and hint not only at the uneven power relations associated with blood sports and patterns of landownership in the Scottish Highlands, but, as in the case of the maharajah, at Scotland's imperial ties. For example, the "deer forests" of the Scottish Highlands were understood to offer the perfect practice ground for those wishing to take advantage of exotic colonial game, as "deerstalking was felt to perfect the field-craft of those with sights trained on the wildlife jungle or high veldt," and thus Scottish sporting estates were a regular "stopping-off point on the grand tour of many a peripatetic sportsman" and empire builder in training.[36]

Yet it was not just the nobility who called on MacDonald's services. For example, the "stock" section of the stuffing books detail that large numbers of skins of otter, wildcat, fox, and pine marten, as well as numerous species of birds of prey, were being sold to the firm by gamekeepers. This highlights how out-of-season game-keeping practice centered on the maintenance of land for elite blood sports, a large part of which consisted of keeping the estates vermin-free so the lairds and landowners could charge premium rates for their shooting leases when the next season came round. Henry McGhie, curator of zoology at the Manchester Museum, has used the records of the birds submitted in Macpherson's "stuffing books" as a historical data set for examining the persecution of birds of prey by sporting estates and estate workers in the Highlands of Scotland.[37] A total of 767 birds of prey were submitted to the store over its working history between 1912 and 1969, 18 of which were hen harriers. According to McGhie's findings, there is no record of a hen harrier submitted from Reay Forest in or before 1926. However, as we ascertained that it was most likely a gun and not a gamekeeper who killed our hen harrier specimen, it could be that McGhie—whose hen harrier records all seem to relate to the "stock" section of the stuffing books, which were in the main submitted by gamekeepers—missed its entry; it may have been part of a bulk submission from a gun in the "work" section of the stuffing book. Of course, it could also be that our harrier was never entered into the stuffing books or indeed that the label is inaccurate. While there is no way of knowing either way—even detailed study of the stuffing books did not reveal conclusively whether MacDonald merely forgot to enter it therein—what McGhie's study does underline is that our tale is not implausible as hen harrier's were being submitted to the store in and before 1926 and that these birds were being submitted in the winter, or prebreeding, season, corroborating Elliot's thesis that our bird was an "adventurer" from the Outer Hebrides.

Sticking to our version of events, then, the gun must have never reclaimed the bird, perhaps thinking its setting up too expensive or because he had enough sporting mementoes already in the form of stag heads. Whichever the case, it inevitably led to the skin simply being "dressed" using the loose-stuff method and entered into the firm's annual year-end sale and its acquisition by the Hunterian Zoology Museum, which was adding to its collection in the 1920s.

## COLLECTION: SITE OF DISPOSAL?

The harrier specimen was acquired for the Hunterian zoological collections in 1926, when zoological science in Glasgow was settling into state-of-the-art premises. The zoology building at the University of Glasgow had been built in 1923 and was named after the Regius Professor of Zoology at the time, Sir John Graham Kerr (1869–1957).[38] Kerr, best known for his work in embryology and as a pioneer of naval camouflage, had a progressive vision about how zoology was to be taught at the university.[39] The driving force behind the creation of the building's zoology museum, he wanted the department to have its own research collections as well as to have a space for demonstrating and independent study. Advocating morphological and phylogenetic approaches to species description, Kerr sought to amass an extensive collection of samples and study specimens so that detailed morphological descriptions of species development could be carried out on-site by both staff and students. Kerr himself greatly advanced knowledge of the evolution of vertebrates through his research on embryology, which he carried out on samples of the South American lungfish (*Lepidosiren paradoxa*) and the pearly nautilus. The samples had been collected by Kerr on an earlier expedition to Paraguay and are still preserved in the museum collections.[40]

While zoological collections were a key means of studying, recording, and communicating scientific knowledge in the past, increasingly, they have come to occupy a marginal position in contemporary biological research. As the fields of molecular biology and biochemistry began to undermine morphological approaches to the study of evolution in the 1950s and 1960s, "laboratories outgrew museums and herbaria as the premier places of modern science" and were therefore no longer considered as active sites of scientific research.[41] Yet as Kevin Winker, an ornithology curator at the University of Alaska Museum, has recently warned: "we cannot be lulled into a view that the day of the collection is past."[42] While he concedes that the collecting mission of the heyday of natural history—to document species diversity and distribution—is widely considered to have been accomplished, he also stresses that the traditional uses of biological collections in taxonomy, systematics, and evolutionary biology account for only part of these collections' value. He demonstrates that, contrary to opinion, collections-based biological research is actually "blossoming" thanks to new RNA and DNA extraction techniques that have helped to extend the usefulness of specimens.[43] The problem, however, is that while those close to collections remain connected to these important developments, to others, as Winker himself recognizes, these collections seem "increasingly arcane."[44]

Yet where some have found only stilled life, it is important to emphasize that there is *still* life, that zoological collections and their specimens are important resources not only for investigating animal life—biology—but also for exploring human-animal relations. For example, the Hunterian Zoology Museum collections, as the Graham Kerr collections are now known, continue to play an important role in the biological teaching, learning, and research, ensuring that the collections are routinely used and added to. The current curator has even welcomed the expanded use of the collections beyond the life sciences, such as the work by Foster. Employing the freedom of an artist to work metaphorically, Foster's interventions in the afterlives of zoological specimens use environmental issues as points of departure, pulling out complexity and paradoxes in issues that otherwise tend to become polarized as they enter political arenas. For example, in *Disposition* (2003; see fig. 3), Foster used the female specimen of a hen harrier, which has been the focus of this essay, to explore the *ongoing* persecution of birds of prey in the interests of game shooting.

Instructed by reports by the RSPB that hen harrier numbers continued to be kept low because of illegal killing and disturbance associated with areas where moors are managed for grouse shooting, Foster aimed to explore and highlight the issue using the Hunterian's harrier specimen as her starting point. The first aspect of the work involved taking the specimen to the estate where it had been shot: Reay Forest. Foster took a series of experimental photographs of the hen harrier specimen at different points in this estate, and exhibited a photograph taken exactly on the point now marked on maps as "Reay Forest." The map mistakenly located its label on a spruce plantation that is now a relatively small part of the Reay Forest Estate. She chose this "shot" even though hen harriers cannot actually nest in forest, as she thought this particular formal composition opened up the most interesting set of associations and questions about changing patterns of land use in the Highlands.[45] Prior to the spruce forest being planted in the 1950s, it would have been part of the "deer forest" and therefore open moorland, a female harrier's favored nesting ground. The second component of work exhibited in the Hunterian Zoology Museum involved redisplaying the bird (which is already camouflaged by its plumage for ground nesting) on a swatch of tweed donated by Westminster estates. This referenced the fact that traditionally, guns and ghillies wore tweed specifically designed to match the plant cover on individual Highland estates, thereby improving their own camouflage. The tweed and study skin were exhibited within a standard glass museum display case, along with a Victorian taxidermy manual whose gold-embossed cover illustrated the method by which the skin of a dead bird of prey might be manipulated into a position of flight.[46]

FIG. 3. *Disposition*. A two-part installation by Kate Foster in the Hunterian Zoology Museum and Department of Environmental and Evolutionary Biology, University of Glasgow, 2003: *a*—photograph taken in a conifer plantation in Reay Forest, Sutherland; *b*—detail of installation in a museum case, showing the hen harrier study skin placed on a swatch of Westminster tweed; *c*—overview of museum display case converted to an art installation within the museum. (Image copyright Kate Foster/The Hunterian Museum and Art Gallery, University of Glasgow)

For Foster, the hen harrier's value, when choosing how to re-represent it in the Hunterian museum on a bed of tweed, was not simply as scientific study skin or discarded dead thing, but rather more as troubling remainder. In such instances of "unfinished or unmanaged disposal," Hetherington argues that the agency of the absent can be expressed through the idea of haunting.[47] While commonly we understand haunting to denote a figural absence, it can also open our eyes to the persistence of presences that remain, and it is for this reason that haunting can take the form of "an unacknowledged debt," explaining why we can feel "a sense of guilt in its presence."[48] In *Disposition* (2003), Foster harnessed the harrier's ability to provoke such feelings not only to highlight the ongoing persecution of hen harriers but also to force an audience to question their responsibility to confront or at least acknowledge this behavior. Foster has said in interview that she saw the harrier as "a transitional vehicle to help move between different kinds of knowing, a creature that can subvert our attention, under the skin."[49] By taking the harrier specimen with her on her initial trip to Westminster Estates, for example, she was able to use it as a reference point from which to then explore different viewpoints and meanings surrounding the species and why it has been both hunted and revered in this context. In this way, the whole process became a public artwork rather than just the "finished" installation in the museum setting. This said, Foster wanted the installation to resonate with the kinds of people likely to visit a university zoology museum. Yet, at the same time, by playing with the conventions by which biological research is made public—adding layers for those prepared to engage with complexity and to share her curiosity about entwined human and animal lives—her aim was to "encourage [visitors] out of their comfort zones, single viewpoints and specialist knowledge."[50]

As part of that audience, Patchett and Lorimer, cultural-historical geographers putting together a proposal for doctoral research that would center on the history of taxidermy and zoological collections at the time, were initially inspired by the idea that individual zoological specimens might have potent afterlives worth examining and extending. Yet after a time they, too, felt an unacknowledged debt to the hen harrier, recognizing that its afterlife (like the thousands of other specimens in store) remained unfinished and could be extended through further acts of creative understanding, prompting a return and this rewriting.

## CONCLUSION

Foster's 2003 reworking of the unique history of the hen harrier specimen relied on its context, and also the background knowledge of its targeted audience, to bring out the various references made in its re-presentation. While Foster was dismayed to learn there is ongoing persecution of harriers in some areas, this was in part countered by learning how other landowners have changed priorities of land management to allow space for birds of prey. For example, on her initial visit to Reay Forest, Foster had prepared herself to meet possible resistance to her inquiries into the harrier's history, yet she actually found a welcome and interest in encouraging breeding in the area from both Elliot (the retired ghillie) and those still working and managing the estate. Yet the situation at Reay Forest sits in sharp contrast to the current practices of estate owners and gamekeepers in southern Scotland and the Pennines, the areas where, unfortunately, harriers have chosen to nest and are in direct competition with grouse.[51] The RSPB reports that in these areas gamekeepers still kill harriers on instructions from high up and only get minor fines if caught, an event that hardly ever happens.[52] While these wildlife crimes in the main go unrecorded, that the numbers of harriers are not rising to what the moorland should be able to support tells the unofficial story.[53]

It is therefore critical to make use of what evidence does remain, in whatever form it takes, to piece together these unofficial stories.[54] This said, and following the aim of this volume, in the case of a hen harrier the real challenge is knowing very much at all about the individual animal, given the fact that human action is aimed at the elimination of material remains and therefore biographical possibilities. While we have been lucky that our harrier skin is a study specimen and therefore offers a form of postmortem biographical data to go on, the question still remains: How well can we ever be expected to (get to) know an animal beyond the postmortem human/animal encounter when addressing zoological specimens? Marvin has stated that usually such specimens "do not begin to have a recoverable history until their final fatal encounter with humans," and thus insinuates that investigators can only really focus on recovering animal *after*lives.[55]

This has largely been the case with our study of the Hunterian harrier specimen given the fact that, as Lorimer noted at the outset, it is difficult to give a description of hen harriers *apart* from humans. Yet, as Winker demonstrates, it is now routine in collections-based research to use historical specimens to conduct retrospective studies. Specimens, according to Winker, "document life in three dimensions: geographic space (locality), biodiversity space (taxonomy), and time (date)," and accordingly they can be understood as "biological filter paper," documenting "experiments" in the environments in which these animals lived." Winker argues that historical specimens should therefore be recognized as a critical data source for reconstructing the past life histories of birds and their shared environments. He also urges that the continued acquisition of new bird specimens should be seen as a priority: "There is clear a need to collect, prepare, and archive specimens in a way that increases the array of preserved components (i.e. animal parts), sample sizes, and dimensions (in biodiversity, geographic, and temporal spaces) available to present and future researchers." If collections continue to be developed in this way, Winker suggests that it is not too outlandish to suppose "that interdisciplinary teams (e.g. ornithologists, entomologists, parasitologists, virologists, isotope ecologists, computational and systems biologists, and community geneticists) will one day delve into a treasure trove of preserved avian stomach and tissue samples to extract complex network analyses of environments, communities, and biospheres."[56]

Here Winker in effect describes an intradisciplinary model for the study of biogeography as it is traditionally defined.[57] The success of this type of collections-based research, however, depends on three factors. First, these collections must be well curated and maintained, which requires a commitment to support and train curators and conservators and to maintain modern facilities. Second, the possibility of new dimensions of study in collections-based research depends on the continued acquisition of specimens and that the data associated with these specimens be as complete as possible. And third, to maximize the usefulness of these collections to researchers, the rate at which this information is entered in databases and made accessible must also increase. Yet, as Andrew Suarez and Neil Tsutsui detail, many biological collections, particularly those associated with museums and academic institutions, have recently experienced reduced financial support for the curatorial work that is necessary for the survival and utility of these collections.[58] Thus, until this disconnect between the possibilities for collections-based research and the financial support that is provided for it is addressed, Winker and other curators may draw resource from our version of biogeographical study.

Our biogeographical perspective, which aims to make do with what remains following an ethic of resourcefulness, makes it possible to make use of smaller collections with more limited data sets, like at the Hunterian.[59] For example, by enacting a form of "salvage ethnography" that works *with* absence and incompletion, even individual specimens with limited data, like our hen harrier, can be utilized as *resources* for telling complex histories of human-animal encounter, cohabitation, and estrangement. While our multifaceted accounts of partial, dispersed, and overlooked remains in "field," "workshop," and "collection" that tell of such histories may disappoint some, our biogeographical approach, which tolerates untidy endings and complexity, also offers points of reconnection. In this case, it has offered ways to look at the present through the past; to make use of a skydancer plucked from the sky by sport, turned inside out, and set in rigid repose for science. Moreover, Foster's highlighting of the harrier's plight through her intervention in its afterlife in *Disposition* (2003) undoubtedly led to two hen harrier taxidermy specimens being included in the newly refurbished Hunterian's display on threatened species. Linking back to Hetherington's idea of attaining settlement with an item's remaining value through disposal, it could be argued that such an act has gone some way toward honoring a debt to a bird whose afterlife otherwise still persists displaced and out of time.

NOTES

1. Mark Cocker and Richard Mabey, *Birds Britannica* (London: Chatto and Windus, 2005), 125–27.

2. Roger Lovegrove, *Silent Fields: The Long Decline of a Nation's Wildlife* (Oxford: Oxford University Press, 2007).

3. Desmond Nethersole-Thompson, "Observations on Nesting Hen Harriers," *Oologist Record* 13 (1933), reprinted in Nethersole-Thompson, *In Search of Breeding Birds* (Leeds: Peregrine, 1992), 43.

4. Brian Etheridge, "Hen Harrier," in *The Birds of Scotland,* ed. Ron Forrester and Ian Andrews (Aberlady: Scottish Ornithologists' Club, 2007), 460–64; Donald Watson, *The Hen Harrier* (London: Poyser, 1977).

5. Jane Desmond, "Displaying Death, Animating Life: Changing Fictions of 'Liveness' from Taxidermy to Animatronics," in *Representing Animals,* ed. Nigel Rothfels (Indiana: University of Indiana Press, 2002), 159–79; Merle Patchett, "Tracking Tigers: Recovering the Embodied Practices of Taxidermy," *Historical Geography* 36 (2008): 17–39; Patchett and Kate Foster, "Repair Work: Surfacing the Geographies of Dead Animals," *Museum and Society* 6 (2008): 98–122; Rachael Poliquin, "The

Matter and Meaning of Museum Taxidermy," *Museum and Society* 6 (2008): 123–34; Bryndís Snæbjörnsdóttir and Mark Wilson, *Nanoq: Flatout and Bluesome: A Cultural Life of Polar Bears* (London: Black Dog, 2006); Karen Wonders, *Habitat Dioramas: Illusions of Nature in Museums of Natural History* (Stockholm: Almqvist and Wiksell, 1993).

6. The "loose-stuff," or "soft-stuff," method is the method employed by most taxidermists and institutions to produce "cabinet" or "study skins" for study collections (see Pat Morris, "An Historical Review of Bird Taxidermy in Britain," *Archives of Natural History* 20 [1993]: 241–55).

7. Garry Marvin, "Perpetuating Polar Bears: The Cultural Life of Dead Animals," in *Nanoq*, ed. Snæbjörnsdóttir and Wilson, 158.

8. Paul Farber, "The Development of Taxidermy and the History of Ornithology," *Isis* 68 (1977): 550–66.

9. Tom Spencer and Sarah Whatmore have argued that biogeography should move away from mapping patterns of spatial distribution and areal differentiation of the "bio" and instead shift attention to the increasingly promiscuous forms of life—and the anxieties associated with them—that populate today's world in mundane and monstrous ways (see Tom Spencer and Sarah Whatmore, "Bio-geographies: Putting Life Back into the Discipline," *Transactions of the Institute of British Geographers* 26 [2001]: 140).

10. For examples of our joint work, see www.blueantelope.info; Kate Foster and Hayden Lorimer, *A Geography of Blue* (unpublished artists' book, University of Glasgow, 2006); Foster and Lorimer, "Some Reflections on Art-Geography as Collaboration," *Cultural Geographies* 14 (2007): 425–32; and Patchett and Foster, "Repair Work."

11. *Disposition* (2003) was the outcome of public artwork funded by the Scottish Arts Council. Foster reworked the piece as a bookwork in 2009 to show them in a gallery context as part of the "Animal Gaze" art show and conference at Plymouth Art College (see Foster's artist's website, www.meansealevel.net).

12. Patchett, "Tracking Tigers"; Merle Patchett, "Putting Animals on Display: Geographies of Taxidermy Practice" (Ph.D. diss., University of Glasgow, 2009).

13. Kevin Hetherington, "Secondhandedness: Consumption, Disposal, and Absent Presence," *Environment and Planning D: Society and Space* 22 (2004): 157–73, 164.

14. Ibid., 167, 168, 166.

15. For example, in *The Biography of a Lie* (2002), Foster tailor-made a collection of body jewelry for bird specimens at the Hunterian Museum that were nearly made extinct by the plumage trade.

16. The 1954 Protection of Birds Act was widely regarded as ineffective as it merely had the effect of driving persecution underground (see Henry McGhie, "Persecution of Birds of Prey in North Scotland 1913–69 as Evidenced by Taxidermists' Stuffing Books," *Scottish Birds* 20 [1999]: 98–110).

17. Lovegrove details that hen harriers are often "lamped" in their ground roosts

under cover of darkness. Another method is to put snow on the eggs in a nest, ensuring there is no evidence left (Lovegrove, *Silent Fields,* 130).

18. Steve Baker, "What Can Dead Bodies Do?" in *Nanoq,* ed. Snæbjörnsdóttir and Wilson, 154.

19. Samuel J. M. M. Alberti, "Constructing Nature behind Glass," *Museum and Society* 6 (2008): 81.

20. Hetherington, "Secondhandedness," 163.

21. www.meansealevel.net/?q=node/5.

22. "Circus" after hawk from the Greek; "cyaneus" from the Greek for dark blue, referencing the male's pearly grey plumage (Roderick Donald McLeod, *Key to the Names of British Birds* [London: Pitman, 1958]).

23. Samuel J. M. M. Alberti, "Objects and the Museum," *Isis* 96 (2005): 559–71; Tim Barringer and Tom Flynn, eds., *Colonialism and the Object: Empire, Material Culture and the Museum* (London: Routledge 1998); Chris Gosden and Yvonne Marshall, "The Cultural Biography of Objects," *World Archaeology* 31 (1999): 169–78; Janet Hoskins, *Biographical Objects: How Things Tell the Story of People's Lives* (London: Routledge, 1998); Janet Hoskins, "Agency, Biography and Objects," in *Handbook of Material Culture,* ed. Chris Tilley et al. (London: Sage, 2006), 60–73; Igor Kopytoff, "The Cultural Biography of Things," in *The Social Life of Things,* ed. Arjun Appadurai (Cambridge: Cambridge University Press, 1986), 64–91.

24. Sarah Whatmore, "Materialist Returns: Practicing Cultural Geography in and for a More-Than-Human World," *Cultural Geographies* 13 (2006): 600–609.

25. Patchett and Foster have previously argued that taxidermy and zoological specimens should be viewed as active assemblages of the movements, materials, and practices which brought them into [and maintain their] existence (Patchett and Foster, "Repair Work," 101).

26. Hayden Lorimer, "Herding Memories of Humans and Animals," *Environment and Planning D: Society and Space* 24 (2006): 515.

27. On the relationship between the financial difficulties of sheep farming and the development of the Highlands as a sporting playground during the Victorian period, see Grant Jarvie and Lorna Jackson, "Deer Forests, Sporting Estates and the Aristocracy," *Sports Historian* 18 (1998): 28.

28. By 1912, 3,599,744 acres, or one-fifth of the entire Scottish landmass, had been converted so that the "gentlefolk" could engage in one-sided mortal battle with the stag, salmon, and grouse (Hayden Lorimer, "Guns, Game and the Grandee: The Cultural Politics of Deer Stalking in the Highlands," *Ecumene* 7 [2000]: 403–31).

29. These conversations consist of a visit to Elliot's house on the Westminster estates by Foster in 2003 and a follow-up visit by Foster and Patchett in August 2009.

30. Edward Harrison, *Scottish Estate Tweeds* (Elgin: Johnstons of Elgin, 1995).

31. In a survey entitled "The 30 Wealthiest Landowners in Scotland 1996," the Duke of Westminster polled top, with 95,100 acres (Jarvie and Jackson, "Deer Forests," 40).

32. See Gordon E. Mingay, ed., *The Victorian Countryside,* 2nd ed., 2 vols. (1981; London: Routledge, 2000), 2:484.

33. Caitlin DeSilvey, "Salvage Memory: Constellating Material Histories on a Hardscrabble Homestead," *Cultural Geographies* 14 (2007): 401.

34. Ibid., 413, 417.

35. While the interview was lost, the museum still had a transcript of the interview to which we were able to refer.

36. Lorimer, "Guns, Game and the Grandee," 416, 414.

37. McGhie, "Persecution of Birds of Prey."

38. John Graham Kerr, "The New Natural History Building," *Museums Journal* 22 (1922): 33–35; Kerr, *Remarks upon the Zoological Collection of the University of Glasgow* (Glasgow: Smith, 1910).

39. Although best known for his studies of the embryology of lungfishes, he also published many textbooks, including John Graham Kerr, *A Textbook of Embryology with the Exception of Mammalia* (London: Macmillan, 1914–19).

40. John Graham Kerr, *A Naturalist in the Gran Chaco* (Cambridge: Cambridge University Press, 1950).

41. Robert Kohler, *Landscapes and Labscapes: Exploring the Lab-Field Border in Biology* (Chicago: University of Chicago Press 2002), 3.

42. Kevin Winker, "Bird Collections: Development and Use of a Scientific Resource," *Auk* 122 (2005): 966.

43. For example, stored tissue specimens from sooty mangabeys (*Cercocebus torquatus*) at the Smithsonian Institution from the late 1800s were used to determine that SIVsm (a simian immunodeficiency virus and a close relative of HIV-2 in humans) was prevalent in Africa at least as early as 1896 (see Andrew Suarez and Neil Tsutui, "The Value of Museum Collections for Research and Society," *BioScience* 54 [2004]: 66–74).

44. Winker, "Bird Collections," 966; Kevin Winker, "Natural History Museums in a Postbiodiversity Era," *BioScience* 54 (2004): 455–59.

45. The spruce plantation, for example, references yet another regretful chapter in the history of land use in the Scottish Highlands. This type of monoculture plantation was hailed in the 1950s as a means of economic regeneration, but, certainly in the far north of Scotland, this ambition was far from fulfilled (see Lorimer, "Guns, Game and the Grandee").

46. Montague Browne, *Practical Taxidermy: A Manual of Instruction to the Amateur in Collecting, Preserving and Setting up Natural History Specimens of All Kinds* (London: Upcott Gill 1878).

47. Hetherington, "Secondhandedness"; Avery Gordon, *Deathly Matters* (Minneapolis: University of Minnesota Press, 1997); Jacques Derrida, *Spectres of Marx* (London: Routledge 1994).

48. Hetherington, "Secondhandedness," 170.

49. Bryndís Snæbjörnsdóttir and Mark Wilson, "Interview with Kate Foster," *Art & Research* 4 (2011).

50. Ibid.

51. For authoritative reports on this situation, see the Scottish Office: www.scotland .gov.uk/library/documents-w/rrrctc-05.htm; or, from the RSPB: www.rspb.org.uk/ ourwork/policy/wildbirdslaw/wildbirdcrime/birdsofprey.asp.

52. Natural England, "The Future of the Hen Harrier in England?" (2008), www .naturalengland.org.uk/publications.

53. Ibid.

54. For example, simply by using records of hen harriers submitted to Macpherson's Sporting Stores, McGhie was able to chart the recolonization of mainland Scotland by this species in the 1940s (McGhie, "Persecution of Birds of Prey").

55. Marvin, "Perpetuating Polar Bears," 157.

56. Winker, "Bird Collections," 967, 970, 968.

57. Spencer and Whatmore, "Bio-geographies."

58. Suarez and Tsutui, "The Value of Museum Collections," 66–67. The main exception to the rule in the United Kingdom is the newly refurbished Darwin Centre at the Natural History Museum, London, which provides world-class storage and research center for the museum's collections.

59. Lorimer's development, and Patchett and Foster's adoption of a "make-do" method, see Hayden Lorimer, "Caught in the Nick of Time: Archives and Fieldwork," in *The SAGE Handbook of Qualitative Geography*, ed. Dydia DeLyser et al. (London: SAGE, 2010), 248–73; and Patchett and Foster, "Repair Work."

HANNAH PADDON

# Biological Objects and "Mascotism"

## The Life and Times of Alfred the Gorilla

B iological collections have been assembled and displayed for centuries in private homes and institutions, public galleries and museums. Contemporary collections are often an amalgamation of historic rare, extinct, common, local, and exotic specimens. These specimens record the changes and revolutions in our knowledge of nature and the environment, our outlook on collecting, and the prestige attached to the collections themselves. They may also trace changes in education and entertainment and reflect societal, political, or cultural values.[1] Like others in this book, this essay considers the multilayered meanings and different values that are, and can be, applied to one particular biological specimen: Alfred the gorilla. His pre-museum life is explored to illustrate how this underpins his success as a taxidermic mount in Bristol City Museum and Art Gallery. In doing so, I want to discuss the concept of "mascotism" as a way of understanding the animal-human connections Alfred generates in his post-mortem life as a prized biological specimen in the museum. Later, I explore the differences in meanings and values between Alfred and other museum mascots before concluding with a wider discussion of mascotism, meaning-making, and object biography.

Like Knut the polar bear (Berlin Zoo), recently deceased, and Lonesome George the Galapagos giant tortoise, Alfred gained notoriety as one of the biggest and most animated animal celebrities of his time.[2] Residing for the majority of his life at the Bristol Zoo in England, he gained enormous popularity. Alfred, however, was not born into captivity. His life began in the depths of the tropical rainforest in the West African country of Belgian Congo (now the Democratic Republic of the Congo), where he was filmed playing in the streets of Mbalmayo, Cameroon, by a group on expedition from the American Museum of Natural History.[3] There, the expedition was told of Alfred's story by his then keeper, a Greek merchant. The baby gorilla was supposedly found clinging to his dead mother, who had been shot by a farmer during a "raid" on the farmer's field. Suckled by a local woman (a fact that was verified after his arrival in Europe), he survived into infancy and was eventually acquired by Bristol Zoo for a staggering £350.

His rise to fame began on his arrival at the zoo in 1930. Named after his benefactor, Sir Alfred Moseley, he was slated as "the only gorilla in captivity in Europe."[4] Alfred drew hoards of admirers who traveled from far and wide to visit him at the zoo. His cage was placed in a prominent position, inside one of the main entrances, for all to see. Amassing more fans and celebrity, Alfred would celebrate his birthday (the day he arrived at the zoo) in style with a birthday cake enjoyed with his chimpanzee friend. Moreover, due to the variable British climate, Alfred was anthropomorphized further by his wearing of wool sweaters, sleeping in his own cot, and being taken for walks around the zoo. As the authors Mullan and Marvin posit, "for many visitors to the zoo, the only way to appreciate, understand or feel for the animals is to impute to them human characteristics."[5] To this end, these appreciations, understandings, and feelings for animals transcend both life and death, a point to which I return when considering the animal postmortem.

Alfred had considerable appeal; he was mischievous, frivolous, and, at times, amusingly nonchalant. He was notorious for his dislike of airplanes, double-decker buses, and bearded men, and was not averse to throwing items from his cage at visiting members of the public. Alfred's tenacity, propensity for troublemaking, and larger-than-life character were complemented by his tenderness and care for the things he cherished: his keeper, Bert Jones; and the sparrows that frequented his cage to feed on breadcrumbs. At the height of his popularity, sales of more than twenty thousand "Alfred" postcards were recorded annually by the zoo.[6] Alfred helped to curb, and ultimately vanquish, the cultural and societal misconceptions of the gorilla painted by the likes of Paul Du Chaillu, who proclaimed the animal to be a "hellish dream creature," and Reverend John Leighton Wilson, who described the species as "one of the most frightful animals in the world."[7]

Reminiscing about childhood visits to the zoo to see Alfred, Mrs. Kemsley recalled in a recent letter to Bristol City Museum and Art Gallery:

> There was something very special, even charismatic about this huge, powerful animal, and even now, after all these years I remember those dark, gentle, thoughtful eyes, mirroring his pride and confidence. In my childish way I felt sure there was an empathy between us, that he knew how much I loved him and each year as I returned, and ran eagerly to renew my acquaintance, I thought that he was as pleased as I that we were together again. [Visiting Alfred at the museum] I stood so long just gazing at my old friend, and my eyes misted over as the years rolled away. Once again I was a shy little seven year old in a crisp cotton frock, black patent shoes and fresh white socks and we were together.[8]

There was a great disparity between Alfred the gorilla—fearsome, uncontrollable, the "hellish dream creature"—and Alfred the beloved, gentle friend of all those who felt they "knew" him.

On 10 March 1948, after twenty-six years at the Bristol Zoo, the *Western Daily Press* and *Bristol Mirror* reported that Alfred, Bristol's much-loved gorilla, had passed away the previous day. In their reports, they claimed that Alfred had died of a heart attack brought on by the low-flying airplanes he so vehemently despised. In fact, Alfred had been struggling with his health for quite some time; he had eventually succumbed to tuberculosis at the grand old age of twenty-eight.[9] His death deeply saddened the people of Bristol, so much so that they couldn't bear to be parted from their beloved ape.[10] As the remainder of this essay reveals, Alfred is now "living" a different existence; his presence in the World Wildlife Gallery at Bristol City Museum and Art Gallery is a continuance of his "life" following his physical death.

## POSTMORTEM: ALFRED AND BRISTOL CITY MUSEUM AND ART GALLERY

Immediately following Alfred's death, his skin was sent to the most famous company of taxidermists in Britain, Rowland Ward. At their studios in London, he was immortalized in an imposing all-fours stance and subsequently returned to the city that idolized him. Placed in his own glass case, Alfred was proudly exhibited in the rear hall on the ground floor of the museum. His close proximity to a highly frequented part of the museum, the café, and his legendary status were somewhat underplayed by his diminutive handwritten calligraphic label. The label, according to Ray Barnett, the museum's head of collections, detailed only basic information. Although Alfred's interpretation was extremely limited, his accessibility and prominence in the museum's hall coupled with his celebrity made him a well-visited and much-loved object.

In 1988, Alfred's position within the museum changed. In a major redisplay and reinterpretation of the existing galleries of biological material, Alfred was removed from his highly visible location on the ground floor to a rather quiet and darkened corner of the World Wildlife Gallery on the first floor of the museum. This move was somewhat contentious, placing him among a plethora of foreign animals in a lone alcove of the gallery. Barnett explained, "We were slightly worried when we moved him upstairs into the new gallery because we thought he might have been hidden, which is partly why we have the Alfred-style directions throughout the building."[11] In his less conspicuous location within the museum and surrounded by other foreign animals, Alfred is certainly facing fiercer competition for visitor attention.[12] For those visitors who do stop to look, there are two levels of interpretation: his biographical narrative is recounted in a short documentary film, while a scientific/conservation narrative stresses Alfred's legacy as a representative of his species and as the longest-lived captive gorilla in Europe at the time of his death.

Recently a case had been made for Alfred's removal from display at the City Museum to allow for his reinterpretation and integration into the displays currently being developed for the new Museum of Bristol in the dock area of the city. Although this was mooted with enthusiasm, it was eventually decided that Alfred should remain in his postmortem home at the City Museum. His death mask, however, never before seen by the public, will be displayed at the new museum as a compromise. Even though Alfred's continued display at the

City Museum has been secured, the museum team are considering his position in the next redisplay of the biological collections. Using multilayered interpretations, the team hope to include new film archives such as *Who Stuffed
Alfred the Gorilla?*—the award-winning short-film spoof by Tom Kelpie
—and a short snippet of sepia film recovered from the American Museum of
Natural History archives in New York.[13] The footage, which was captured
by an expedition team during the early 1930s, shows Alfred playing with two
other orphan gorillas in the streets of Mbalmayo. There are plans to expand
not only on his personal story but also his wider legacy. The new displays, it is
hoped, will concentrate on the plight of the gorilla species and use Alfred as
a platform to discuss habitat conservation, climate change, extinction, and so
on in a gallery where he would take center stage.[14]

Although Alfred's afterlife may seem rather humdrum, he has, in fact, been
the focus of many campaigns and certainly one conspicuous event. Sparking a
major police investigation in the summer of 1956, Alfred was stolen from the
City Museum in the dead of night by three university undergraduate students
as part of a RAG-week stunt (charity "Raising and Giving" events are held annually in many UK universities). The exact details of the notorious heist have
only recently come to light following the death of one of the perpetrators, who
had believed that the revelations would lead to legal action.[15] As reported by
the press in 2010, the students gained entry to the museum having had a key
cut. Sneaking into the gallery in the early hours of the morning, they removed
Alfred from his case and transported him to a studio apartment in Bristol,
where he was dressed in various guises and remained for three days. Alfred
was safely returned to the museum, much to the relief of staff and the general
public, via a doctor's office waiting room, where he waited unabashedly, much
to the bewilderment of the office caretaker. But why had Alfred been selected
as the participant of a well-thought-out RAG-week stunt? Another of the three
perpetrators claimed, "We took Alfred because he was such a big Bristol personality."[16] Needless to say, Alfred has not been beyond the museum walls
since the stunt.

Alfred has also been the inspiration for numerous events and activities during his display at the museum. In the 1970s, Alfred was the muse for a play
by native Bristolian Peter Nichol, *Born in the Gardens,* which debuted at the
Theatre Royal in Bristol from August to September 1979. Alfred was at the
heart of the story, a lighthearted comedy about "caged inmates" Maud and

Maurice, and "escapees" Queenie and Hedley.[17] In 1993, Alfred became the focus of another event. The museum dedicated to Alfred a week of short talks, given by curators and Bristol Zoo staff, which examined his life at the zoo, his character, and his ambassadorial role for the gorilla species. During the week-long event in June, visitors to the museum were invited to write in, sharing their memories of and stories about Alfred.[18] In doing so, the museum added to the wealth of biographical material accumulated over the years since his accession, including anecdotal stories of mischievous behavior and spellbound audiences.

Furthermore, in 2008, Nick Jones and Toby Lucas made the short documentary that now plays next to Alfred. In an interview for the film, Ian Redmond, chief consultant for the Great Apes Survival Partnership, touched upon the importance of Alfred's role in changing people's perceptions of the gorilla: "For the first time people in this area were able to see a male, adult gorilla and wonder at what an amazing being the gorilla is. [They were encouraged to] think of them not as creatures or monstrous animals but as non-human beings."[19] Also in 2008, the one-time "Dictator of Bristol," as Alfred was sometimes called (a jibe at the European fascist leaders of the period), was transformed into a cartoon figure as part of a "family-friendly campaign" launched by the museum.[20] Consultation with the museum's family audience revealed that Alfred was by far the true "mascot" of Bristol City Museum and Art Gallery.[21] Today, freestanding advertising boards emblazoned with Alfred's image grace the museum's entrance, inviting passersby into the museum to discover the collections on display (fig. 1). In addition, Alfred's cartoon image appears in various guises as he promotes the city's numerous museums and historic houses. He also heads up the new museum trails with his taxidermied skin strategically signposted for those eager to visit him. The museum has continued to promote Alfred as a mascot and icon for both the museum and the city.

In his pre-museum and postmortem lives, Alfred has garnered admiration, awe, and affectation. He has helped to change the perceptions of the gorilla species and created countless memories for those who knew him in life and those who now know him in death. He is highly prized and remains within the consciousness of Bristolians today. Having discussed his life and afterlife, I focus in the second part of this essay on Alfred, and similar museum animals, as "mascots," particular objects for meaning-making.

FIG. I. Alfred as mascot:
Bristol City Museum and
Art Gallery family-friendly
marketing campaign. (Bris-
tol Museums, Galleries and
Archives)

## MEANING-MAKING AND MASCOTS

Over the last decade, museologists have explored the concept of meaning-
making, a "sociocultural approach to mind-growing" in the museum context
which considers the ways individuals learn about and make sense of the objects
and messages displayed in the museum.[22] By applying their own experiences,
knowledge, and values, visitors will inevitably develop unique perspectives
and ideas about an object's meaning. This concept, Hein argues, is a "phenomenon of nature (not just a theoretical construction)"—nor is it limited to
the museum. Another concept, particularly pertinent to Alfred the gorilla as
he sits in Bristol City Museum and Art Gallery, is mascotism—he is "a thing
used to symbolise a particular event or organisation; an emblem."[23] Other

examples include Sir Roger the elephant (Kelvingrove Art Gallery and Museum; see Richard Sutcliffe, Mike Rutherford, and Jeanne Robinson's essay in this volume), Rosie the rhino (Ipswich Museum), and Buddie the lion (Paisley Museum and Art Gallery). Alfred's use as a museum mascot can helpfully be explored in relation to meaning-making, particularly considering his ability to create connective experiences, invoke memories, and tell a unique story.

Biological specimens in museums can be singled out as mascots for a number of reasons. First, a recent survey of British museums provided evidence that the majority of mascots are large, exotic mammals, although there are some exceptions, for example, Winkie the pigeon (McManus Galleries, Dundee).[24] Second, mascots tend to be taxidermied, although again there are exceptions to the rule, for example, Maharajah the elephant (Manchester Museum, see Samuel Alberti's essay in this volume) and the Baron of Buchlyvie (the skeleton of a Clydesdale horse at Kelvingrove Art Gallery and Museum). Third, these animals may have acquired fame or "significance" in their pre-museum lives, much like Alfred, Chi-Chi the panda (Natural History Museum, London; see Henry Nicholls's essay in this volume) and Wallace the lion (Saffron Walden Museum). Finally, most mascots are anthropomorphized through naming, either during life or in their afterlife at the museum. For instance, Buddie the lion was named following a competition at the Paisley Museum and Art Gallery. Assistant Curator of Natural History Nicola Macintyre explained, "In 1999 to celebrate his 70th birthday the museum ran a children's naming competition in which he was given the name Buddie" (people from the Scottish town of Paisley are also generally known as "Buddies").[25] For the purpose of this essay, mascotism is simply defined as those specimens of biology on display which have an elevated meaning—whatever it may be—for visitors and museum staff alike.

So, what is it that makes Alfred, and other mascots, special? The answers may well lie within the biography of the animal. An object biography, or life story, "prioritises the inextricable link between things and people by focusing on the meanings constructed around objects."[26] Igor Kopytoff suggests that these constructed meanings are plentiful and unique as objects move through their "lives."[27] Ivo Maroević concurs with the idea of multiple biographies but refers to them as "multilayered identities."[28] These layers, elucidated by Peter van Mensch as conceptual, factual, actual, functional, and structural identities, are important as they provide extra dimensions in which to interpret an object.[29] Perhaps, in light of these explanations, object biography seems the most obvious form for interpretation and meaning-making available to the museum professional. Object biographies are often revealed through labels, display panels, and gallery videos. How visitors perceive, interpret, and relate to these stories, or constructed meanings, may differ greatly from the intended messages emitted from the museum.[30] Roger Silverstone asserts that the biological object biography in the museum presents the opportunity for the continuation of "the imaginative work of the visitor who brings to it his or her own agenda, experiences and feelings."[31] Curators, education officers, and designers can play to the strengths of the object's histories and interpret them in meaningful ways. And this is certainly the case for Alfred. The label on the glass case in which he is exhibited narrates a life story, littered with heart-wrenching and heartwarming tales of his survival, his new home in Britain, and his eventual demise.

His multilayered, complex biography can be likened to other animal celebrities' life stories. One of the most notorious biographies is that of Wallace the lion, who was taxidermied in the early 1830s and has remained on display at Saffron Walden Museum since 1835. His story began in Edinburgh, where he was thought to be the first lion bred in captivity in Britain. As part of Wombwell's Travelling Menagerie, Wallace became infamous for a well-documented and particularly fierce fight with dogs at Warwick.[32] It is also thought that he was the inspiration for Marriott Edgar's poem "The Lion and Albert."[33] Although Wallace represented a different social, cultural, and political period in British history, he also gained a celebrity akin to Alfred's, and his story is still told in the museum today.

Socially and culturally, Alfred challenged the negative perceptions associated with his species. Although his appearance was imposing and at times ferocious, he was gentle-natured. Scientifically, he was studied in order to advance human knowledge of the gorilla, the largest of all the world's primates. Records of his diet (Alfred was a strict vegetarian) and illnesses suffered were kept alongside details of his weight and stature.[34] Politically, the title afforded to Alfred during the 1930s and 1940s, the "Dictator of Bristol," reflected the rise of fascism in Europe. Animal celebrities like Alfred were also integral to the promotion of the conservation movement and new conservation organizations in the 1960s and 1970s such as the World Wildlife Fund and Greenpeace (see Henry Nicholls's essay in this volume).[35]

Like other mascots, Alfred was and is anthropomorphized in two inter-related ways.[36] The first is the anthropomorphic journey the museum object takes prior to and upon entering the museum collections; the second, the anthropomorphizing of the object either before or after death. The former may be from their encounter with the collectors who identify, classify, and display them, and/or from the curators, researchers, educators, conservators, and visitors who have different forms and varying degrees of contact with them.[37] One particular layer, which all biological specimens acquire, is their human-imposed, scientific identity. In attempting to systematize nonhuman beings, humans have regulated the animal kingdom with the introduction of the binomial nomenclature. This classifying of biological objects adds to the existing range of social, physical, cultural, and political biographical layers.

Alfred and others mascots are further anthropomorphized when they are individually named. Greg Mitman discusses how 1960s animal behavior stud-ies involved the naming of individual animals, thus transforming the non-human creatures from objective to subjective beings with human attributes and emotions.[38] The pre- and postmortem biographies and attributed values and meanings of museum mascots can be influenced by their acquiring a hu-manized name. Naming a biological object in the museum display context can also enhance the status and importance of an otherwise "normal" specimen, including, for example, Gerald the giraffe (Royal Albert Memorial Museum, Exeter). For many of the animal mascots in British museums, names were ac-quired during their lifetime in captivity, including Alfred, Wallace, Chi-Chi, and Guy the gorilla (at the Natural History Museum and briefly displayed at Weston Park Museum, Sheffield). Mullan and Marvin argue that "certain people use anthropomorphic devices in order to understand animals and con-struct animal identities."[39] Indisputably, this includes the naming of the ani-mal, which often transcends the evolution from pre-museum, living creature to postmortem object. In Alfred's case, Mullan and Marvin's "anthropomor-phic devices" are further recognized in his celebrating his "birthday" with par-ties at the zoo and wearing wool sweaters to allay the harsh British climate.[40]

But why do humans continue to place anthropomorphic, and anthro-pocentric, ideals upon animals both pre- and postmortem? Mary Midgley takes the view that "we are not just rather like animals; we are animals." "Our differences from other species may be striking," she continues, "but compari-sons with them have always been and must be, crucial to our view of ourselves."[41] A rather more anthropocentric position is taken by Lorraine Daston and Gregg Mitman, who proclaim "humans project their own thoughts and feelings onto other animal species because they egotistically believe themselves to be the centre of the universe."[42] Whatever our connection with animals, it seems we are prone to applying anthropomorphic perspectives onto nonhuman beings. I believe, however, that it is those imposed anthropomorphic constructions that connect audiences with animals, and museum visitors with animal mascots. Furthermore, I would suggest that in the societal psyche, animal celebrities such as Alfred can be placed near the center of a linear continuum between human and animal. They are not regarded merely as animals, but neither are they human. They acquire humanistic attributes in life, thanks to the humans around them, and these often continue postmortem.

One final link between meaning-making and mascotism is memory. In the museum, memory "evokes emotions and desires, positively or negatively charged; memory is also driven by a desire to remember or forget."[43] Objects can reawaken memories of experiences from childhood, adolescence, and adulthood. They can place unlimited value upon observed objects and provide opportunities for new meaning-making. In a brief study of mascots in British museums, memory emerged as an outcome of visitor interaction with these beloved animals. Responses about mascots asserted that they "reminded people of their childhoods, or when they first visited the museum," but they were also capable of creating new memories: memory making.[44]

John Falk and Lynn Dierking report that museum visits can be vividly recalled in later years.[45] In fact, memories of museum experiences and specific objects can not only be recalled years later but can "trigger . . . an endless range of personal associations."[46] In 1993, the Bristol City Museum and Art Gallery ran a competition which invited members of the public to write into the museum with memories of Alfred at the zoo. They received countless recollections of Alfred encounters from visitors (including Mrs. Kemsley's, quoted above). These encounters between the visitor and Alfred have personal meanings; for example, a perceived friendship between a young child and a wild animal. Similarly, countless relationships between Alfred and the museum visitor have developed since his inception as a departed icon in the museum. His presence in the museum has ensured that he continues to build new memories for those who observe him.

Not only can encounters with mascots reawaken distant memories and trigger personal associations, audiences may also encounter these museum objects on another level. Kiersten Latham explains that some visitors may have a "numinous experience" that is "a deeply, connective, transcendental encounter . . . with a museum object."[47] I suggest that in Alfred's case his numinous radiance has been directly transferred from his life at the zoo to his afterlife in the museum. He still has the power to mesmerize an audience and evoke strong memories, creating a sense of awe and wonderment. Further, objects can stimulate collective memory, a "cultural memory that speaks beyond individual experience."[48]

A representative of the changing attitudes toward his species, Alfred was a champion for gorillas in Britain at the time of his arrival to Bristol. By the time of his death, he had gained the collective adoration of not just the citizens of Bristol but of millions of people across the United Kingdom and beyond. His position as museum mascot projects the collective memory from days past to the present. Many thousands of visitors pass by his cabinet each year and, with the help of his label, reawaken the collective cultural memories he induces. Mascots remain significant to museum audiences even after long periods of time.

To complement Alfred's story of creating collective memories, it is helpful to turn our attention to Gerald the giraffe, displayed in his own dedicated gallery at the Royal Albert Memorial Museum since the early Edwardian period.[49] With a recent, successful bid to the Heritage Lottery Fund, the museum has won funding to redevelop its galleries. In doing so, the project team have taken the bold step of relocating Gerald to another gallery. The marketing manager for the museum explained: "If we said we were getting rid of Gerald there would be uproar. . . . You talk to anybody about the development [and] one of the first questions they ask you is 'what's happening to Gerald?'"[50] The power of these objects to transcend the museum itself, to project their meanings onto audiences so much so that the audience demands to be kept informed about their beloved mascot, is remarkable. The belief that the removal of the mascot from display would provoke such strong remonstrations from museumgoers was echoed by the curator, who proclaimed: "The giraffe is such an iconic object that I don't think anybody would have been able to say 'Right, let's not have the giraffe on display. We're tired of seeing it.'"[51] This considered relocation of Gerald has impacted on the subsequent redisplay of the entire biological collections at the Royal Albert Memorial Museum. Gerald is no longer simply being displayed as the museum's mascot; his story is being told through his extensive biography in a gallery which focuses on how objects were amassed by the museum. Undoubtedly Alfred would provoke similar interest and remonstration.

Further qualitative research unearthed visitor feelings about another museum mascot, Snowy the polar bear of Weston Park Museum in Sheffield. As the museum redisplayed its biological collections, Nick Dodd, chief executive of the Sheffield Museums and Galleries Trust, reported that during front-end evaluation, the focus groups told the museum redisplay team "in no uncertain terms that if we got rid of the polar bear, they wouldn't be coming back."[52] From the research, it is therefore reasonable to assume that any mascot would provoke such strong reactions from their audiences if they were to be removed from display (or simply moved within the museum). The importance of the museum to the local community can help build an identity, something which people can relate to or in which they can take pride. Because, of course, for the thousands of visitors who regularly venture to the museum, their mascot has become familiar. Familiarity in the museum promotes a sense of security and continuity. In fact, it "is perhaps the easiest method for allowing visitors to make connections to museum exhibitions."[53]

While Alfred is classed within an elite group of biological museum objects, mascots, he was also—like Guy the gorilla, Chi-Chi the panda, Wallace the lion, Knut the polar bear, and Lonesome George—an animal celebrity during his life. As Michelle Henning observes in her essay in this volume, the scientific, societal, cultural, political, and economic identities of these animals are inextricably linked; humans make connections and bring animals closer. Our anthropocentric views of the animal kingdom and the anthropomorphic ideals we apply contribute to layered identities, constructing celebrities with fascinating stories.

Alfred is by no means, then, the only zoo animal to gain celebrity. He is not the only gorilla exhibited in a museum. He is not particularly large and does not have any distinguishing features. Yet his life story has captured the hearts of generations. He has challenged and helped change attitudes toward the gorilla. He was, and still is, an ambassador for his species, which is now listed as critically endangered by the International Union for Conservation of Nature and Natural Resources.[54] He is visited by children and adults daily; he tells his story and will continue to tell his story; he will never be removed from display. He is a true mascot.

## NOTES

1. Carla Yanni, *Nature's Museums: Victorian Science and the Architecture of Display* (Baltimore: Johns Hopkins University Press, 1999); Stephen. T. Asma, *Stuffed Animals and Pickled Heads: The Culture and Evolution of Natural History Museums* (Oxford: Oxford University Press, 2001); Tony Bennett, *The Birth of the Museum* (London: Routledge, 1995).

2. www.knut.net; Henry Nicholls, *Lonesome George: The Life and Loves of a Conservation Icon* (Basingstoke: Macmillan, 2007).

3. Ray Barnett, "The Dictator of Bristol," *Nonesuch: The University of Bristol Magazine,* Spring 1999, 38–40.

4. Bristol Zoological Gardens, *Bristol Zoo: Book and Guide* (Bristol: Bristol Zoological Gardens, n.d.), 17–18; "Alfred Heads the Popularity List," *Western Daily Press,* 5 August 1933.

5. Bob Mullan and Garry Marvin, *Zoo Culture: The Book about Watching People Watch Animals,* 2nd ed. (Urbana: University of Illinois Press, 1999).

6. Richard Boston, "Alfred, Bert, Mum and Me," *Weekend Guardian,* 23 June 1990, 25.

7. Paul Belloni Du Chaillu, *Explorations and Adventures in Equatorial Africa* (London: Murray, 1861), 71; John Leighton Wilson, *Western Africa* (New York: Harpers, 1856).

8. Abstract from a letter received by Bristol City Museum and Art Gallery on 15 June 1993 from an entrant, Mrs. Kemsley, to the "Memories of Alfred" competition, Bristol City Museum and Art Gallery files.

9. Barnett, "The Dictator of Bristol," 40.

10. "The Story of Alfred the Great," *Evening Post,* 30 July 1992, 18.

11. As discussed with Ray Barnett, head of collections for Bristol City Museums, 20 January 2010.

12. Graham Black, *The Engaging Museum: Developing Museums for Visitor Involvement* (Abingdon: Routledge, 2001), 201.

13 Tom Kelpie, *Who Stuffed Alfred the Gorilla?* short-film spoof, 2008, www.youtube.com/watch?v=ZXeB7PvTrDU.

14. As discussed with Ray Barnett, 20 January 2010.

15. Luke Salkeld, "Revealed after 54 Years: Secret of Alfred the Ape's Abduction in Rag Week," *Daily Mail,* 5 March 2010, 3.

16. Ibid.

17. Charles Spencer, "Born in the Gardens: An Irresistibly Quirky Comedy," *Telegraph*, 24 July 2008.

18. As discussed with Ray Barnett, 11 February 2009.

19. Nick Jones and Toby Lucas, *Alfred the Gorilla*, film installed in the World Wildlife Gallery at Bristol City Museum and Art Gallery in 2008. Jones and Lucas are award-winning filmmakers from the University of the West of England, Bristol.

20. As discussed with Rhian Rowson, acting curator of Biology at the Bristol City Museum and Art Gallery, 8 July 2008.

21. As discussed with Claire Royall, publicity officer for Bristol City Museums, 20 January 2010.

22. Shawn Rowe, "The Role of Objects in Action, Distributed Meaning-Making," in *Perspectives on Object-Centred Learning in Museums,* ed. Scott G. Paris (Mahwah, N.J.: Lawrence Erlbaum, 2002), 19–36; Josh Gutwill-Wise and Sue Allen, "Finding Significance: Testing Methods for Encouraging Meaning-Making in a Science Museum," *Current Trends in Audience Research and Evaluation* 15 (2002): 5–11.

23. Mascot (noun), in *Oxford English Dictionary* online, http://dictionary.oed.com/.

24. Hannah Paddon, "An Investigation of the Key Factors and Processes That Underlie the Contemporary Display of Biological Collections in British Museums" (Ph.D. diss., Bournemouth University, 2010).

25. Nicola Macintyre, assistant curator of Natural History at Paisley Museum, e-mail message to the author, 1 December 2008.

26. Susan Langdon, "Beyond the Grave: Biographies from Early Greece," *American Journal of Archaeology* 105 (2001): 579–606.

27. Igor Kopytoff, "The Cultural Biography of Things: Commoditization as Process," in *The Social Life of Things: Commodities in Cultural Perspective,* ed. Arjun Appadurai, 64–94 (Cambridge: Cambridge University Press, 1986).

28. Ivo Maroević, "The Museum Message: Between the Document and Information," in *Museum, Media, Message,* ed. Eilean Hooper-Greenhill, 200–214 (London: Routledge, 1995).

29. Peter van Mensch, "Museology as a Scientific Basis for the Museum Profession," in *Professionalising the Muses: The Museum Profession in Motion,* ed. Van Mensch, 85–95 (Amsterdam: AHA, 1989).

30. Eilean Hooper-Greenhill, *Museums and the Interpretation of Visual Culture* (London: Routledge, 2000).

31. Roger Silverstone, "The Medium Is the Museum: On Objects and Logics in Times and Spaces," in *Towards the Museum of the Future: New European Perspectives,* ed. Roger S. Miles and Laura Zavala (London: Routledge, 1994), 164.

32. For description and illustration of Wallace's fight with Tinker and Ball, the baited dogs, see Pierce Egan, *Anecdotes of the Turf, the Chase, the Ring, and the Stage:*

*The Whole Forming a Complete Panorama of the Sporting World* (London: Knight and Lacey, 1827). Wallace the lion was the inspiration for Marriott Edgar's poem "The Lion and Albert." For Wallace's intriguing biography, see *Ravishing Beasts,* www.ravishingbeasts.com/lions/2006/11/3/wallace-the-lion.html.

33. See *Ravishing Beasts,* www.ravishingbeasts.com/lions/2006/11/3/wallace-the -lion.html.

34. Barnett, "The Dictator of Bristol," 39.

35. For the growth and development of the WWF, see Peter Denton, *The World Wildlife Fund* (New York: New Discovery Books, 1995); for the history of Greenpeace, see Michael Brown and John May, *The Greenpeace Story* (London: DK, 1989).

36. For definition of anthropomorphism, see Elliott Sober, "Comparative Psychology Meets Evolutionary Biology," in *Thinking with Animals: New Perspectives on Anthropomorphism,* ed. Lorraine Daston and Gregg Mitman (New York: Columbia University Press, 2005), 85–99.

37. Samuel J. M. M. Alberti, "Objects and the Museum," *Isis* 96 (2005): 559–71.

38. Gregg Mitman, "Pachyderm Personalities: The Media of Science, Politics and Conservation," in *Thinking with Animals,* ed. Daston and Mitman, 175–95.

39. Mullan and Marvin, *Zoo Culture,* 1.

40. Barnett, "The Dictator of Bristol," 39.

41. Mary Midgley, *Beast and Man: The Roots of Human Nature* (London: Methuen, 1979).

42. Lorraine Daston and Gregg Mitman, "The How and Why of Thinking with Animals," in *Thinking with Animals,* ed. Daston and Mitman, 4.

43. Susan A. Crane, Introduction to *Museums and Memory,* ed. Crane (Palo Alto: Stanford University Press, 2000), 1.

44. Paddon, "An Investigation of the Key Factors and Processes."

45. John Howard Falk and Lynn Diane Dierking, *The Museum Experience* (Washington, D.C.: Howells House, 1992).

46. Gaynor Kavanagh, *Dream Spaces: Memory and the Museum* (London: Leicester University Press, 2000), 3.

47. Kiersten F. Latham, "The Poetry of the Museum: A Holistic Model of Numinous Museum Experiences," *Museum Management and Curatorship,* 22 (2007): 247–63.

48. Marius Kwint, "Introduction: The Physical Past," in *Material Memories,* ed. Kwint, Christopher Breward, and Jeremy Aynsley, 1–16 (Oxford: Berg, 1999).

49. James R. Ryan, "Hunting with the Camera: Photography, Wildlife and Colonialism in Africa," in *Animal Spaces, Beastly Places: New Geographies of Human-Animal Relations,* ed. Chris Philo and Chris Wilbert, 203–21 (London: Routledge, 2000).

50. As discussed with Ruth Randall, marketing manager for the Royal Albert Memorial Museum, 22 July 2008.

51. As discussed with David Bolton, curator of natural history at the Royal Albert Memorial Museum, 15 November 2007.

52. Nick Dodd quoted in Alfred Hickling, "Bring on the Giant Spiders," *Guardian,* 1 May 2008.

53. Hein, "Learning in the Museum," 161.

54. International Union for Conservation of Nature and Natural Resources, Red List of Threatened Species, Version 2010.1, www.iucnredlist.org.

MICHELLE HENNING

# Neurath's Whale

In 1933, the American magazine *Survey Graphic* published an article entitled "Museums of the Future" by the Viennese museum director and polymath Otto Neurath. Neurath gave the example of a typical whale exhibit to explain what he saw as the limits of natural history displays:

> A huge whale hangs in the middle of the hall; but we do not learn how the "beard" is transformed into old-fashioned corsets, how the skin is transformed into shoes, or the fat into soap that finds its way to the dressing room of a beautiful woman. Nor do we learn how many whales are caught per annum, or how much whale-bone, fat and leather are produced by this means. And yet many people surely would be interested to know what this means for the balance of trade, how it relates to economic crises, and so on. Human fortunes are connected with this exhibit—starving seamen, hungry families of fishermen in the north of Norway. And so, everything leads to men and society.[1]

Neurath's whale is a hypothetical, not a particular whale. He invokes it to question the separation between human society and nature that museum displays reinforced and that was rooted in the disciplinary divide between the humanities and the natural sciences.[2] His view was that the museum's purpose was to help the visitor to understand his or her own place in social processes. In 1933, whaling was still big business, and Neurath proposes that the current dependence on whales be made visible so that the animals are revealed to be part of industrial modernity, transformed into commodities and abstracted as stocks and shares. It is a vivid and interesting example of Neurath's museological approach, and one of the few times he describes an object-based museum display in his writing.

Yet it is not clear exactly what kind of object Neurath meant by "a huge whale." He may have been thinking of the whale skeletons exhibited in the Naturhistorischesmuseum in Vienna, or in the central hall of Berlin's Museum für Naturkunde, or that he had seen on visits to Britain and the United States. Few examples of cetacean taxidermy exist since dead whales putrefy quickly and whale skin is difficult to preserve, so models were increasingly used in exhibits. Neurath's essay predates the blue-whale model at the British Museum (Natural History) that was completed in 1938, but in the United States there was the 76–foot-long papier-mâché model of a blue whale, suspended since 1907 at the American Museum of Natural History (AMNH) in New York.

This hypothetical whale is not entirely imagined, then, but based on the experience of visiting whale halls and seeing suspended specimens or models. As well as challenging disciplinary boundaries and the separation of nature and society, Neurath is questioning the continued orientation of museums toward the "quasi-religious" display of singular, impressive objects.[3] Neurath proposes a reorientation: the whale may remain at the center of the hall, but the real center of the exhibit becomes "man and society" rather than the whale. This raises interesting questions about the museological display of whales, and about the relationship between objects and contextual information in displays. Could this huge object simply slip into the background, quietly allowing us to move from the contemplation of it to the contemplation of the human uses of whales? What, in the context of an exhibit about whales' social uses, would visitors see? Would the whale even remain visible as "a whale"? And, if it could succeed, would Neurath's proposed exhibit connect us more closely with the whale, as he seems to suggest?

Consider what Neurath wants the exhibit to do. He wants it to make visible a "network" of human and nonhuman relationships and dependencies, and to overcome the existing splitting of nature from society. Yet, in the case of whale exhibits, museum natural history was already closely entwined with industry. Exhibits that featured whaling were prepared by museums, including the

Smithsonian, which produced fishery exhibits for a succession of international expositions in the 1870s and 1880s. These exhibitions were so frequent that the museum secretary, Spencer Baird, complained that the task of producing them was hindering the museum's usual work. Baird's combined roles as secretary of the Smithsonian, director of the United States National Museum, and commissioner of Fish and Fisheries suggest that political-industrial interests and scientific concerns were very much entangled.[4]

In the Smithsonian exhibit for the 1883 Great International Fishery Exposition in London, scientific and museological concerns mixed with the vested interests and propagandistic purposes of industry and state. The exhibit promoted national identity and progress, and aggrandized U.S. industry, including whaling. Before it opened, the *New York Times* had predicted that it would explain "the gradual march of the fisheries" in the United States to the present day, in which "the value of our catch exceeded that of any other country." The *Times* noted that the exhibit concerned itself not only with industry but also with the scientific classification of fish (and whales). It would also include some of the things Neurath later described as absent from the whale hall: harpoons, guns, and other weapons used in hunting were to be shown but also the "archaeology of whaling," alongside a perfectly equipped whaling boat, whaling logbooks, and "all the makeshifts of the whaler"—"such as strange lamps manufactured out of tumblers" and "rough skates fashioned out of files." In addition, there was a life-size model of a harpooner standing on the bow of a whaling ship. The article concluded that this exhibit would certainly be "in the highest degree creditable to our country."[5] Indeed, the Americans came away from the 1883 exposition with eighteen gold medals.

Neurath's aim was politically very different from the propagandistic aims of the International Exposition. He set out to facilitate democratic participation in social planning, which meant first that museums were an educational tool to help visitors understand their relationship with a wider world. The museum should aim to make knowledge relevant and coherent, a process he later described as the "humanization" of knowledge.[6] In Neurath's imagined exhibit, the aim is not to aggrandize whaling, but to show the interdependence of man and animal. Neurath's text proposes bringing natural history home, making the whale not simply an object of curiosity and wonder or scientific interest but something of direct personal concern to visitors. Thus, Neurath makes the whale social and political.

However, the Smithsonian exhibits do point to a problem with Neurath's critique. For Neurath, it is the separation (of whales from us) enacted by natural history that is the dilemma. For the sociologist Bruno Latour, separation is only part of the problem. In his book *We Have Never Been Modern*, Latour describes "the modern divide between the natural world and the social world," but he is also interested in the ways in which the divide is continually overcome and reinstated. Through a process he calls "translation," modern science produces "entirely new types of beings, hybrids of nature and culture." Another process, "purification," then reinstates the separation: "two distinct ontological zones, that of human beings on the one hand, that of nonhumans on the other."[7]

The fact that the 1883 Smithsonian exhibit included whales (mammals) as fish demonstrates the priority of social, industrial classifications over scientific ones, yet at the same time, science and the museum lend the exhibit authority. Even the whaleman appears as an anthropological subject through his ingenious inventions. Scientific knowledge and industrial interests are simultaneously mixed (translated) and kept apart (purified), since rigor, disinterestedness, and objectivity assure scientific authority. The whale appears caught up in human activity, part of culture, and yet the translation of the whale into a social being is also denied through purification since whales are reduced to "nature" in the form of raw material to be exploited by humans. The whale is pushed out, assigned its place as nonhuman, in the same moment that it seems to be being accepted as part of the social. The same could be said for Neurath's whale, which also appears primarily as a human resource.

Nevertheless, Neurath's envisaged whale exhibit also invokes beauty, hunger, and starvation. The dismembered whale connects peoples across the globe, connects the desire for beauty with others' hunger. If "everything comes back to man," it is not to rational, scientific man, but to the diversity of feeling, living, eating humans. Human attachment to the whale, human dependence on the whale is not "happiness neutral," to use Neurath's own phrase.[8] At the same time, the feelings evoked here all relate to the use of the whale and to the hunt.

We can read Neurath's explanation of what is missing from the whale hall as indicative of what whales were to people at the time. The whale was industrial material, often described in terms of the number of barrels of oil it yielded. Philip Hoare's book *Leviathan: Or, The Whale*, inspired by *Moby-Dick*, shows vividly how, at the height of the industry, and for those directly involved, the whale was known primarily as a calculable, disassembled object.[9] Whalemen and whalers looked at the animal with trained eyes: they estimated and described its size and strength with a view to its capture, in relation to volumes of oil and blubber, with an eye on market prices.

Yet the whale is also a fearsome creature, and the encounter with a living whale is an encounter with another conscious being which orients itself toward you, which responds, acts, and reacts. Encounters with these creatures would have evoked strong feelings, however much the whalemen were trained to see them as potential commodities, or in terms of the work they would undertake on their corpses. They saw the whale as an industrial object, labored on the whale's body, and also could be inspired to terror and awe as they faced the creatures they slaughtered. As the literary theorist Philip Armstrong expresses it, nineteenth-century whalemen had an experience that "routinely alternated between dangerous encounters with the vast materiality of the living animal and its reduction to dead and partial resources, a commodity to be measured by the barrel, reified by the factory ship's technological procedures and its specialization of labor."[10]

This vacillation between the two perspectives may have diminished by the time Neurath was writing, as the industrialized mass extermination of whales was then under way. New technologies and forms of industrial organization meant an end to the struggle between man and whale, and perhaps to the sense of the whale as an awe-inspiring "monster." Whales were slaughtered on sight and en masse, and the huge corpses were processed on factory ships in ever-more-efficient ways, weighed, measured, and reduced to saleable units with no chance for either side to step back and wonder at the other. Whales were now a war resource, for use in the manufacture of nitroglycerin. As John Berger's famous essay "Why Look at Animals?" suggests, modern industrial society sets animals at a distance from us, making it hard to see the animal as codependent, to relate to it as more than an object.[11]

Berger's essay also suggests that this detachment of humans from other animals allows symbolic, emotive, and sentimental attachments to animals to proliferate rather than diminish. The whale remained, and is still, a powerful creature in myth and symbolism. We do not need any actual encounter with a whale, life-size whale model, or whale skeleton to develop strong, even passionate, feelings toward it. The possibility of love for the whale, of fear, empathy, awe may be provoked by narrative accounts, by fictions, rumors, pictures, and our capacity to imagine. These feelings are variable and historically specific. Few people now face a whale knowing they have to kill it or may be killed by it. Moreover, I cannot say whether the tears that come to my eyes when I see film footage of a whale's tail rising and sinking below the surface of the ocean might have any of the same emotive content as the tears in a nineteenth-century sailor's eyes when he saw the same sight for real. A gulf lies between us, in which a photographic cliché has been born, whales have been driven to the edge of extinction, and environmentalism has become a popular cause. Even so, in the 1930s as now, most people's relationship with whales was with imagined whales. These do not possess the same certainty, the same calculable quality as "the balance of trade" and "economic crises," but they affect us just as much.

## THE PROPERTIES OF THE WHALE

Although he ignores other emotional and meaningful attachments to the whale, Neurath is not arguing that we need see the whale solely in instrumental terms. He simply suggests we restore the social character to something that has been misconceived, and misrepresented, as purely nature. This is consistent with the Marxist perspective that Neurath knew well. Marx used the example of a cherry tree to argue that objects we conceive of as "nature" are the product of human activity. Commerce enabled the cherry tree to appear before us as an immediate, natural thing. As Marx wrote, "the sensuous world is not a thing given from all eternity, ever remaining the same, but the product of industry and the state of society . . . a human product."[12] Similarly for Neurath, "everything leads to men and society." In his proposed exhibit, there is therefore no sense of the whale as a creature existing regardless of, and outside, human society and human work on it. Clearly the whale as an animal does exist regardless of us. Alive, it is a beast with its own purposes, its own intentions, its own unfathomable world. To encounter a living whale is, as Berger suggests, to encounter something that looks back, but across a gulf: it is to encounter an "other," similar and different from oneself.[13] Yet once the whale is disassembled into commodities or transformed into a museum exhibit, it is, like Marx's cherry tree, the product of human labor, social transactions.

One way in which the whale exhibit becomes a whale for us is through the imposition of ideas and beliefs, which then appear to us as apparently innate properties of the whale itself. We impose meaning onto the things of the world by bringing a certain "baggage" to any encounter. For example, certain kinds of illustrations and written encounters of whales may have predisposed the nineteenth-century museum visitor to view the whale as a "monster." Museum models or specimens are never neutrally exhibited. They do not appear before us as simple matters of fact, but are produced by various exhibition practices, contextual framings, institutional conventions. To see the whale as something that is socially and culturally constructed means attending to how such practices, frames, and institutions actually *produce* the object as a whale. Various discourses or scripts make it possible to see and act toward the display in one way and not another. The whale-object is not simply sensuously given to us, nor does it appear to a viewer innocent of the discourses and context that surround it. The ways we perceive, represent, and classify the whale are rooted in our own culture.[14]

This is complicated, however, by the fact that visitors encounter it bodily, as a material thing, in actual space, and via their senses. From a phenomenological perspective, we might say that the whale (object) and the visitor (subject) work together.[15] The museum visitor orients her attention, and also herself, toward it. The object impresses itself upon her materially, but the properties she perceives as belonging to it are not simply "in" the object but in its relation to her. If the whale model has certain material properties that she takes into account, this is because of her particular bodily engagement with it: thus, weight becomes a felt quality of the whale skeleton only for those people assigned the task of hanging it in the hall. The workers doing this will experience it as resistant or malleable, as workable or not, just as whale oil materially lends itself to certain industrial uses and not others. Every subjective encounter with the object, even for visitors just looking, is also a material encounter. By exhibiting the whale in certain ways and allowing visitors to engage in certain ways (to touch, or not to touch, for example), an exhibition might cross out certain properties of the whale while directing visitors toward others.

Text, labels, and other contextual material inform how visitors see the whale model or skeleton. Other informative exhibits—photographs and charts, animals of related species, and other whale remains—orient visitors toward the whale in the center of the hall, directing them to these properties and not those. Certain properties of the whale exhibit become vivid, while others recede as irrelevant, go out of focus, or disappear altogether. In this sense, changing the exhibition context as Neurath proposes would change the whale itself.

This process of recontextualization, by which some of the properties of an exhibited object become perceptible and others disappear from view, may be described as the object's "social life." The notion of the "social life of things" is usually attributed to the anthropologists Arjun Appadurai and Igor Kopytoff, and emphasizes changeability. A good example from the art historian Philip Fisher concerns a hypothetical sword that changes hands and moves from being a weapon through to being treasure and loot, and eventually becoming a museum specimen. In the museum, the sword is no longer lifted and swung, and is now an object of primarily visual and narrative, educational interest. Fisher shows how objects shape and constrain human activity and how, through their place in a "community of objects," they direct our attention, but he also argues that the traits of an object are only real within certain social scripts: "our access assembles and disassembles what the object is." Referring to his sword example, he writes: "Once it is bolted down in a display and not swung in a certain way we cannot say that balance or imbalance is even a fact about it. Without a class of warriors, trained to fight in certain ways, even the permission to lift and swing the sword could tell us nothing."[16]

The whale has certain traits, certain properties that result from its material existence in the world. But, on this account, these matter only if visitors can access them. In altering our access, orientation, and interpretation, exhibitions alter the properties of objects. This does not mean that all entities are socially constructed; very few social scientists or cultural theorists would argue that. Instead, they tend to retain a category of the natural, or extrasocial, for certain kinds of objects. Or, as Latour expresses it, we tend to divide "nature" into two kinds of things: "soft" objects that are "white screens" for projection (that is, what we see as their natural properties are actually our own ideas projected onto them), and "hard" larger forces (such as biology, or laws of physics) that shape society. However, Latour argues against this division. Either we need to see things like the laws of physics as the arbitrary projections of society, he writes, or we have to rethink the idea that any objects are simply white screens. He introduces the term "quasi-objects" to describe those things that are "much more social, much more fabricated, much more collective" than the hard forces or objects, and yet they are not arbitrary projections, but "much more real, nonhuman and objective."[17]

If the whale hall were to include objects or text communicating the information Neurath suggests, and if this were to make "everything lead to man and society," the whale in the hall would have to be a "soft" object that can simply have new meanings attached to it. If we take this view, we treat the suspended whale as a white screen, one of those objects that are "mere receptacles for human categories."[18] Even in Fisher's example, it could be argued that swords and other objects have social lives only insofar as they are instruments for people, enlivened through human attention and uses. By contrast, Latour gives a much stronger meaning to a thing's liveliness, suggesting that material things confront us with their materiality, demand from us certain kinds of care, refuse to do some things, differentiate between us, and do other things that were never intended by us. This does not make them alive, but it does make them "actants," not as active as "agents" but not simply playing a prescribed role, either. Things participate in our social lives, not simply as a means to our ends, but alongside us. We are "enveloped, entangled, surrounded" by things, passionately connected and interdependent. Latour conceives of them as "complex assemblies of contradictory issues" and as material, actual, substantive.[19]

Latour also suggests that the semiotic and material properties of a thing cannot be separated. The whale exhibit's communicative aspect is part of its materiality, not a superimposed layer that we can peel off and replace at whim. Misled by the "modernist opposition between what was social, symbolic, subjective, lived and what was material, real, objective and factual," we have tended to forget that symbols are always material and to treat the material as if this were just a medium, a carrier.[20] Suspended in its great hall, the whale is a designed object, whether model or reconstructed skeleton, put together to represent a whale, and therefore like all representations, to draw attention to certain aspects, to invite us to think about whales in certain ways. It is the product of a process of construction (the measurements, designs, the manipulation of papier-mâché and wood). It speaks to us (if ambiguously) through its physical construction. Even the skeleton is produced as a signifying object. A process has been gone through to make it stand for a whale: the bones have been cleaned and processed, joined together, made into a construction that can then be hoisted up to the ceiling. Through these procedures they are made to describe or depict a whale.

Yet a skeleton also conceals this, seeming to say, "I am whale," rather than, "this is what the whale is like." I have written elsewhere about how taxidermy, like photography, stakes its truth claim in the fact that it both looks like and is made out of the animal it represents.[21] Taxidermy demands that we respond to it as more than just an object. Thus children ask their stumped parents, "Is it real?" and "Is it dead?" In the case of skeletons, there is less of the visible form of the animal present, but there is still a claim to realism: we assume this is what the animal looks like "underneath" or "inside." Models, however, do not have this direct relationship to the thing; they have no status as scientific specimens and are purely display objects.

In the case of the massive blue-whale models mentioned earlier (at the Smithsonian, the AMNH, and the Natural History Museum in London), accuracy is very difficult for visitors to assess, since they are highly unlikely to have seen living blue whales. Even Richard Van Gelder, the chairman of the Department of Mammology at the AMNH who was asked to oversee the production of a new blue whale in 1959, had never seen one.[22] Nevertheless, according to the science historian Michael Rossi, the model was based on thorough studies, with Van Gelder "insisting, for instance, that twenty-eight tiny hairs be placed in the massive model's chin, in accordance with what was known about blue whale whiskers."[23]

If measurements and photographs have been taken, if various procedures have been adhered to, the scientific character of the display might be assured, and models can also count as truthful, reliable, and authentic displays.[24] The 1907 whale at the AMNH was made of wood, iron, and papier-mâché, but it was based on photographs of dead whales and on the Smithsonian's whale model, which itself was based on plaster casts laboriously taken from a dead whale.[25] The "mechanical objectivity" attributable to photography and the directness of the plaster cast underwrite the model's truth claim, but so do the expert knowledge and skills of the paleontologists, taxidermists, and biologists who produced the models. In other words, the whale-object becomes a convincing whale not only through direct physical connection with the animal it represents, but through the rigor of the processes involved and the authority of science. As Lorraine Daston and Peter Galison argue in relation to scientific illustration, objectivity in science is developed through such techniques and practices, regimes and routines.[26] The exhibit is the result of certain technical procedures, rituals, habits, disciplines, and ideologies.

Rossi points out that the 1907 AMNH model was "by no means a generic model" but referred to a specific animal and "preserved the zero-sum relationship between referent and representation that underwrote the other taxidermies in the museum."[27] However, the nineteenth-century commitment to truth to nature meant that the specific singular whale had also to be generalized, to become the typical whale, by eliminating or at least playing down the anomalous or the atypical aspects of a specimen. The whale-object, whether skeleton or model, is always already, to adopt another word from Latour, an "imbroglio"—a tangled knot of human social practices and whalishness, of material, technical thingliness, and ideas. Neurath's hypothetical whale would not be a bounded solid object, but a multiple, entangled quasi-object—already socialized, already embedded in the world of natural science and its ways of seeing the world but also concretely present before the museum's visitors. Because it is already so, we cannot say that it has no properties, or that its properties are dependent on our access.

It is hard to imagine a whale-model as an imbroglio. As Hoare reminds us, the earth had been pictured from space before the first underwater photographs of whales were taken: "We knew what the world looked like before we knew what whales looked like."[28] Whales, that is, fully in their element. The whale models in museums preceded this, giving a sense of the whale's suspended grace, but the analogy is a good one, because somehow whales (especially blue whales) seem as bounded and as whole as the earth does in those pictures. Before that, large whales were mostly perceived in fragments, as parts that did not add up. Those parts that surfaced were allegedly taken for islands or for other fabulous beasts. Caught, or beached, whales become formless, ungainly things. Yet in representation they seem bounded, self-evident, and also singular—"the whale." So much so that when the philosopher Ian Hacking writes (in *The Social Construction of What?*) of how categories of people get produced as "a definite class" or subspecies in the singular ("the child viewer," "the woman refugee," "the disabled individual"), he says, "like 'the whale.'"[29] He could have chosen any animal, but the whale is the most appropriate.

So the whale model is already embedded within, and has embedded within it, practices, procedures, and protocols, yet it appears to us as whole and unassailable because of its shape and its sheer size. The blue-whale model suspended from the ceiling of the Natural History Museum in London is designed to impress me with its scale. This is obviously a material property but also a designed, symbolizing, meaningful aspect of it. To fully comprehend this aspect, it helps if I can get my body close to the (model) whale body. Since size is relational, my experience of the whale's largeness, and my comparative

smallness, is between the two of us. No other objects need join this community, no beautifully designed exhibits need point me to this; the main thing
I need is proximity. I also need some guarantee that this is how a blue whale
looks and this is its true size.

How I then understand and experience my encounter with a museum
whale will depend on all sorts of factors—I may have picked up from other
displays and paintings a Romantic understanding of the relationship between
the human body and the vastness of nature, and that might shape how I feel
about this. In any case, the difference in scale is unlikely to be a neutral fact
for me, and I am unlikely to respond with indifference. The natural history
museums that hit upon the idea of suspending their whales from ceilings well
knew their spectacular and emotive power. By the 1950s and 1960s, whale
exhibits were designed with the explicit intention of inviting visitors to feel as
if they had entered the whale's underwater domain.[30]

## MANAGING WONDER

Size connects the museum experience with the experience of living whales.
Even in contexts where the whale was seen primarily as an industrial-techno-
logical resource and a set of commodities, its physical presence was overpower-
ing and had to be managed. In the nineteenth- and early-twentieth-century
port cities that prospered from the whaling industry, the whale was both a
symbolic and physical presence. Ishmael, the narrator in Melville's *Moby-
Dick,* remarks of the city of New Bedford:

> Nowhere in all America will you find more patrician-like houses; parks
> and gardens more opulent, than in New Bedford. Whence came they?
> how planted upon this once scraggy scoria of a country?
>
> Go and gaze upon the iron emblematical harpoons round yonder
> lofty mansion, and your question will be answered. Yes; all these brave
> houses and flowery gardens came from the Atlantic, Pacific and Indian
> oceans. One and all, they were harpooned and dragged up hither from
> the bottom of the sea.[31]

In this contrast between an airy neighborhood and the violent activity that
made it possible, houses and gardens are shown to be whalish in scale. They
become interchangeable, one by one harpooned—one house, one whale. In
places like New Bedford, the physical presence of the dead whale's body was

concealed in the wealthy parts of the city. The whale was transformed into a symbol, and from 1856, New Bedford's city hall proudly bore the motto *Lucem Diffundo* (We light the world).[32] But the viscera of the whale was less easy to forget in the port itself. Very often the whale bodies were managed by being dissected and processed before the ship arrived in port. Nevertheless, in New Bedford, whale oil had "saturated the soil and the air was redolent with the heavy odor," as one speaker recalled with nostalgia at the 1916 opening of the Bourne Whaling Museum.[33]

By the time the whale's body permeated the mechanisms of modern civilization, it no longer smelled and it had been efficiently reduced to manageable amounts of raw material. It spread into fashion and commodity culture in the form of face creams, corsets, and leathers, into industry as machine lubricants, and lit the streets of the great metropolises. In these places, the visceral presence of the whale could not be felt. Such a transformation took a great deal of labor and industrial effort. Large whales, such as sperm whales, were difficult and dangerous to hunt, and turning their bodies into useable materials was also a dangerous, physical process. The whale corpse was, first of all, a mass that had to be managed with effort (Richard Sabin's essay in this volume makes this very evident).

As Marx explained, human labor is concealed in the commodity, so that the opulent displays of the department stores seem to have nothing to do with factory labor. Similarly, the whalish element was concealed in the whale-based products that appeared on the market (soaps, skin creams, umbrellas, varnish, and margarine, for instance). When whale products were gradually superseded by other materials, such as petroleum, spring steel, rapeseed oil, and eventually plastic, the change was hardly felt, except in those places like New Bedford where it meant unemployment and decline.[34]

Managing whales' vast corpses was part of exhibition culture, too. A very vivid example of the unmanageability of the whale's body after death is given in an 1889 *New York Times* article about "a giant whale that has lately been exhibited in the Prater" (in Vienna). Problems with the preservation of the animal meant that "the monster has not been able to resist the laws of nature and has gradually passed into a state of complete putrefaction," resulting in an "effluvia" that "has pervaded the whole Prater." Intended as a gift to the museum, the whale instead had to be "given over to the public flayer" to be cut up and buried.[35] By Neurath's time, embalming provided a partial solution to this dilemma. Traveling whale shows were big business in the United States from 1928 (though they had existed since around 1920), and they continued until about 1937. The principal touring company was the Pacific Whaling Company, a show-business concern which also toured a whale to Britain in 1931. The Pacific Whaling Company's displays of humpbacks and finbacks were very profitable. For their first display, they acquired a whale from a ship that usually provided whales for the soap industry and attempted to embalm it, but it exploded. Undeterred, they proceeded to exhibit it very profitably, despite its increasingly putrid smell.[36]

To manage the whale's immense body is extraordinarily difficult. People went to such effort because whales were profitable, and they were so because of the demand to experience them firsthand. How this direct experience was assimilated is another question. Attempts to bring the whale closer, to allow it to be seen directly, took place in the context of a society in which fragmentary, compartmentalized information dominated. For example, writing in 1940, the critic Walter Benjamin observed that newspapers prevent readers from assimilating news content into their own experience, and that this separation is perpetuated through the style, layout, and journalistic practices that inform the newspaper and are part of its form.[37] Latour also says of newspapers: "All of culture and all of nature get churned up everyday . . . but the analysts, thinkers, journalists and decision-makers will slice the delicate network for you into tidy compartments where you will find only science, only economy, only social phenomena, only local news, only sentiment, only sex."[38]

Neurath's plan to reorient the classic whale-hall exhibit toward "man and society" is about overcoming such compartmentalization, reconnecting the artificially separated spheres of the whale and human daily experience. This cannot be done by simply bringing the whale closer. It also requires overcoming the arbitrariness of modern education, the lack of connection between topics as well as the lack of relevance to lived experience. He wrote, "The wealth of scientific detail is no longer held together by a unitary approach, and in a certain sense it is left to chance whether a man thinks about some linguistic formations in Chinese or about a medieval text, about African beetles or about wind conditions at the North Pole."[39]

However, there is a distinction between this kind of arbitrary education and the compartmentalization of knowledge that emerges from increased specialization and division of labor. The expert knowledge of whales that circulated in the whaling era was based in the pursuit of whales for profit; it was motivated, instrumental knowledge. Arbitrary knowledge tends to be nonmotivated, and associated with the leisure practices of the aristocracy, the working and lower-middle classes. While an education left to chance could be the product of neglect (a poorly organized museum, a badly designed curriculum), interest led by wonder, whim, and curiosity is also characteristic of the dilettante, the leisured amateur, and of popular entertainments. Some forms of mass entertainment, such the newspaper, seemed to echo the fragmentations and jarring juxtapositions that had become a feature of everyday experience and working life in modern, urban society.[40] Others, though, offered compensatory thrills, escapes from everyday life, and knowledge that was valued precisely because it was not immediately useful.

Whales were most explicitly made into objects of curiosity and wonder in the traveling whale shows mentioned earlier. They combined circus techniques with mortuary skills, scientific information with whalers' firsthand narratives. Whale shows toured via the railroads, and their success meant they were more than a sideline of the whaling industry: preparing dead whales for exhibit became an "assembly line process." More whale morticians were trained and employed. As the public became more accustomed to the spectacle, the Pacific Whaling Company adopted new techniques to draw the crowds. They advertised in newspapers, using the same publicity techniques as the circus and many of the same showmen. The whales were accompanied by lectures and sometimes other animal-based attractions including (at the other end of the scale) flea circuses.[41]

Whales had become sensational, and the chief sensation to be had was the feeling of proximity: to get close to the whale, to touch the creature, even if it had to be dead for this to happen. In 1954, Pathé News filmed the model Anita Woolf climbing in the mouth of the sixty-five-foot embalmed "Jonah the Whale" (fig. 1). Such images glamorized the show's promise of proximity. Mass culture brings things closer, as Benjamin observed with reference to technological reproduction, but reproduction was no substitute for the traveling whale show. Bringing whales closer involved immense effort: the newsreel of Jonah's arrival shows a 100-foot lorry negotiating London streets with difficulty, and a voice-over explains that the whale could not be taken by train because its immense weight threatened to break the rails. Proximity was also assured through the lectures that accompanied the 1930s whale shows. These

FIG. 1. The model Anita Woolf turns to face the camera after climbing into the mouth of the Jonah the Whale in 1954. (British Pathé)

were delivered by seasoned showmen using nautical language and dressed as sailors, convincing the audience they were whalemen who had "felt the mist in their faces as they harpooned the whales they were showing."[42]

In Neurath's time, exhibitions and traveling shows allowed audiences to experience the enormity of the whale: the very quality that makes it difficult to manage also makes it wondrous. So, although Neurath is critical of "arbitrary knowledge," an exhibition which connects whales to its human uses, in an attempt to overcome the separation of knowledge and experience, needs also to allow for the wondrousness of whales (both whales themselves and whale exhibits). As long as the huge whale suspended in the center of the hall remains a thing at which to wonder, it will be relevant to its audience's lived experience, not by leading us toward "man and society," but by leading us away: because wonder might unseat us, allow us to forget ourselves, even—like the women in the photograph and film—to be consumed by the whale.[43]

NOTES

1. Otto Neurath, "Museums of the Future," *Survey Graphic* 22 (1933): 458–63, reprinted in *Otto Neurath: Empiricism and Sociology,* ed. Marie Neurath and Robert S. Cohen (Dordrecht: Reidel 1973), 219–20.

2. Hadwig Kraeutler, *Otto Neurath: Museum and Exhibition Work* (Frankfurt am Main: Peter Lang, 2008), 184.

3. Ibid., 222–23.

4. Dean C. Allard, *Spencer Fullerton Baird and the U.S. Fish Commission* (New York: Arno Press, 1978); Smithsonian Museum website, http://vertebrates.si.edu/fishes/ichthyology_history/exhibits.html.

5. "Vienna's Dead Whale," *New York Times,* 14 July 1889.

6. Otto Neurath, "Visual Education: Humanisation versus Popularisation," unfinished manuscript, 1945, reprinted in Neurath and Cohen, *Otto Neurath,* 231–32.

7. Bruno Latour, *We Have Never Been Modern,* trans. Catherine Porter (Cambridge: Harvard University Press, 1993), 10–13.

8. Otto Neurath, "International Planning for Freedom," *New Commonwealth Quarterly* (April 1942), reprinted in Neurath and Cohen, *Otto Neurath,* 427.

9. Philip Hoare, *Leviathan: Or, the Whale* (London: Fourth Estate, 2008), 31.

10. Philip Armstrong, "'Leviathan Is a Skein of Networks': Translations of Nature and Culture in Moby-Dick," *ELH* 71 (2004): 1039–63.

11. John Berger, "Why Look at Animals?" in *About Looking* (London: Writers and Readers, 1980).

12. Karl Marx, *The German Ideology,* 1975, 70.

13. Berger, "Why Look at Animals?"

14. On discourse, see Herbert L. Dreyfus and Paul Rabinow, *Michel Foucault: Beyond Structuralism and Hermeneutics,* 2nd ed. (Chicago: University of Chicago Press, 1983).

15. Sara Ahmed, *Queer Phenomenology: Orientations, Objects, Others* (Durham, N.C.: Duke University Press, 2006), 55.

16. Philip Fisher, *Making and Effacing Art: Modern American Art in A Culture of Museums* (London: Harvard University Press, 1991), 18–19.

17. Latour, *We Have Never Been Modern,* 52–55.

18. Ibid., 52.

19. Bruno Latour, "A Cautious Prometheus: A Few Steps Toward a Philosophy of Design (With Special Attention to Peter Sloterdijk)," keynote lecture for the "Networks of Design" meeting of the Design History Society, 3 September 2008, 4, 8, www.bruno-latour.fr/articles/2008.html.

20. Ibid., 6.

21. Michelle Henning, "Skins of the Real: Taxidermy and Photography," in *Nanoq: Flatout and Bluesome: A Cultural Life of Polar Bears,* ed. Bryndís Snæbjörnsdóttir and Mark Wilson (London: Black Dog Press, 2006), 25–46.

22. Van Gelder cited in Michael Rossi, "Modeling the Unknown: How to Make a Perfect Whale," *Endeavour* 32, no. 2 (2008): 59.

23. Michael Rossi, "Fabricating Authenticity: Modeling a Whale at the American Museum of Natural History, 1906–1974," *Isis* 101 (2010): 360.

24. Ibid., 341.

25. Ibid., 342.

26. Lorraine Daston and Peter Galison, *Objectivity* (New York: Zone, 2007).

27. Rossi, "Fabricating Authenticity," 343.

28. Hoare, *Leviathan*, 31.

29. Ian Hacking, *The Social Construction of What?* (Cambridge: Harvard University Press, 2000), 10.

30. Rossi cites Van Gelder's memoir on building the whale at the AMNH: "You are a skin diver without apparatus. You are one with the sea" (Rossi, "Fabricating Authenticity," 360).

31. Herman Melville, *Moby-Dick, or The White Whale* (1851; London: Everyman's Library, 1991), 32.

32. James M. Lindgren, "'Let Us Idealize Old Types of Manhood': The New Bedford Whaling Museum, 1903–1941," *New England Quarterly* 72 (June 1999): 167.

33. Ibid., 165.

34. Ibid., 167.

35. "Vienna's Dead Whale," *New York Times*, 14 July 1889.

36. Sam Abbott, "Whales Smelled out the $$," *Billboard*, 28 June 1952, 48, 63, 96. For more on the commercial display of whales, see Fred Pfening, "Moby Dick on Rails," *Bandwagon, the Journal of the Circus History Society* 31 (1987): 14–17.

37. Walter Benjamin, "On Some Motifs in Baudelaire," in *Walter Benjamin: Selected Writings*, vol. 4, *1938–1940*, ed. Howard Eiland and Michael W. Jennings, trans. Edmund Jephcott et al. (Cambridge: Belknap Press of Harvard University Press, 2003), 316.

38. Latour, *We Have Never Been Modern*, 2.

39. Otto Neurath, "Personal Life and Class Struggle" (originally published in German in 1928), in Neurath and Cohen, *Otto Neurath*, 294–95.

40. Walter Benjamin, "On Some Motifs in Baudelaire," 328–29.

41. Abbott, "Whales Smelled out the $$," 96.

42. Ibid.

43. On wonder, see Stephen Greenblatt, "Resonance and Wonder," in *Exhibiting Cultures: The Poetics and Politics of Museum Display*, ed. Ivan Karp and Steven D. Lavine (Washington, D.C.: Smithsonian Institution Press, 1991), 42–56.

HENRY NICHOLLS

# The Afterlife of Chi-Chi

During the 1960s, Chi-Chi the giant panda—London Zoo's most valuable inmate—achieved global superstardom. Born in the wild in 1957, in the mountains of Sichuan Province in China, she was taken to Peking Zoo, Moscow, Berlin, Frankfurt, and Copenhagen, before being purchased, in September 1958, by the Zoological Society of London (or ZSL, with assistance from Granada Television) for £12,000. She was an instant hit with the public, but her fame peaked in 1966, when the Zoological Society of London sent her back to Moscow Zoo to hook up with An-An, then the only other giant panda in captivity outside China. There was no mating. The "Soviet panda" came to London a couple of years later, but again without reproductive success. So by the early 1970s, with Chi-Chi's health beginning to fail, it had become clear she would leave no offspring. Nevertheless, the image of this singular panda has been reproduced in many other ways, with profound consequences for how we see this species and the natural world.

## THE END OF AN ERA

In the six months before Chi-Chi's death on 22 July 1972, "she had suffered a number of convulsive attacks which resembled epilepsy."[1] She was treated with the anticonvulsant pimidone, which did not help matters, and the vitamin B thiamine, which did. But with her increasing reluctance to show herself in public, the zoo's press office began to field phone calls from alarmed admirers fearing for their panda. In April, a presenter for the British Broadcasting Corporation's (BBC) news program *Nationwide* telephoned the zoo for an update on her condition. The ZSL's public relations officer, Tony Dale, was upbeat. "When we last went to look at her she'd had her dish of tea, then retired into her bedroom for an afternoon nap," he reported. "She is asleep now, on her back very happily waving her paws in the air." But Dale was not confident that she would survive the summer. "I don't think we can give any predictions on this because she's the oldest panda we know of in any zoo and fifteen in panda language is a very old lady."[2]

As she continued to decline, preparations were made for her death. Almost a decade earlier, the Chicago anatomist Dwight D. Davis had published the first detailed study of panda anatomy based on the dissection of Su-Lin, the first panda to be brought out of China alive.[3] But Su-Lin was a young male, and Davis had had access only to the "embalmed and injected body." Not only would a postmortem of Chi-Chi reveal the anatomical secrets of an adult female panda, the opportunity to dissect a recently departed specimen opened up some avenues of inquiry that had not been available to Davis. So the London Zoo's curator of mammals, Michael Brambell, began to think about how he might make the most of Chi-Chi's remains.

In mid-July, Chi-Chi declined her food and then, after three months without any convulsions, suddenly suffered a series of six violent fits in just a few hours. Not long after the zoo closed to the public that day—Friday, 21 July—Brambell administered a sedative "on humane grounds to prevent any self-inflicted damage."[4] He stayed with her through the night, but she never came around. Brambell remembered her death at three o'clock on the Saturday morning as "a very sort of stark and cold time of day."[5]

## THE POSTMORTEM

That Sunday's newspapers mourned the passing of a panda that had "won the hearts of millions around the world."[6] By this time, Brambell and a crack team of experts—the ZSL's pathologist, Ian Keymer; the eminent St. Bartholomew's Hospital Medical College anatomist Alexander Cave; and the British Museum (Natural History)'s (BM(NH)) senior taxidermist, Roy Hale—had finished with Chi-Chi's remains.[7]

The first ghoulish task was to remove Chi-Chi's eyeballs from their sockets. Brambell had lined up the right eyeball to travel to the Medical Research

Council Vision Unit at the University of Sussex in Brighton. There, Professor Herbert Dartnall would carry out investigations into the pigments contained in Chi-Chi's retina from which he hoped to discover (among other things) whether pandas have color vision. Working under red light so as not to excite the pigments, he thawed out the macabre specimen, slit it open, and carefully removed the retina. Reporting his findings in *Nature* the following year, Dartnall concluded that the panda has two light-sensitive pigments, one most responsive to red and the other to white light.[8] Here was an indication that the panda could see in color like most carnivores that are active in the day, a finding that has recently been confirmed by experiments on live pandas.[9]

The left eyeball went in a different direction. Brambell dropped it into a fixative solution and had it sent to the Institute of Ophthalmology in London, where Professor Norman Ashton took detailed measurements and wrote up a brief paper on its gross structure.[10] Other parts of Chi-Chi's body provided more insights into panda biology. Her postmortem took up a whole issue of *Transactions of the Zoological Society of London,* with papers on her alimentary tract, nervous system, vitamin D transport, mammary gland secretion, neurochemistry, and cytochrome c.[11] These were gaps in contemporary zoological knowledge that Davis, working with less than fresh panda material, had been unable to fill.

This scientific endeavor saw the dispersal of Chi-Chi's body parts to institutions across London, farther afield in Britain, and even across the Atlantic to the United States. For example, blood and tissue samples were sent to Vincent Sarich at the University of Berkeley in California, who hoped Chi-Chi's proteins would help resolve a long-running scientific controversy. The giant panda's precise position in the tree of life had been hotly contested ever since its formal scientific description in 1869. Opinion was divided between two main camps: those who considered the panda to be most closely related to bears and those who felt it should be grouped with the red panda and the related raccoons.

Davis had reviewed the controversy in his monograph on panda anatomy published in 1964. "Opinion as to the affinities of *Ailuropoda* is divided almost perfectly along geographic lines," he wrote. In short, he was accusing English-speaking researchers of pulling down English-language publications and non-English-speakers of dipping into exclusively non-English publications, with both camps simply rehashing what they had read. For Davis, this could only mean one thing, "that authoritarianism rather than objective analysis has really been the determining factor in deciding the question."[12]

Davis had tried to draw a line under the debate with his new set of anatomical data based on Su-Lin, a study that the evolutionary biologist Stephen J. Gould would later describe as "probably the greatest work of modern evolutionary comparative anatomy."[13] But in spite of his best efforts, the wrangling continued. So when new molecular methods for ordering species began to emerge, it was only natural that molecular biologists should have wanted to apply them to the panda problem. Using an immunological approach to assess the similarity of giant panda, bear, and raccoon proteins, Sarich concluded that the "association of the Giant panda and the other bears is clear and unequivocal."[14] Although this study did not shut down the debate (which rumbled on for at least another decade), Chi-Chi's blood played its own small part in one of the most protracted disputes in the history of taxonomy. What's more, Chi-Chi called it correctly: today, the overwhelming consensus is that the giant panda is indeed a bear.

In addition to Sarich's analysis and the other academic papers that appeared in the *Transactions of the Zoological Society of London,* Chi-Chi's remains achieved something else worth noting: an impressive collaboration. The long history of research into the giant panda is characterized by intense rivalry between individuals, institutions, and even nations. Even today, there is competition over pandas. The captive population in China, for example, is effectively divided in two, with the Ministry of Construction responsible for those animals in zoos and the State Forestry Administration in control of animals at the China Conservation and Research Center for the Giant Panda in the Wolong Nature Reserve. In this context, the large team of experts that Brambell marshaled in 1972 to pore over Chi-Chi's remains is quite remarkable, a truly international effort that in panda circles has only become more commonplace since the late 1990s.

## THE BRITISH MUSEUM (NATURAL HISTORY)

With the postmortem complete, Brambell offered up what remained of Chi-Chi to the BM(NH), now known as the Natural History Museum. It seems like the obvious place for her to go, but this was not always the first-choice destination for the zoo's deceased inmates.

After it was formally founded in 1826, the ZSL set about landscaping Regent's Park to house a collection of live animals to interest and amuse the public. But almost as important was a parallel project to establish a museum on Bruton Street in Mayfair. So when the zoo eventually opened to the public, any of its exotic creatures that passed away went straight to the society's museum rather than to the BM(NH)—a practice that would later be echoed at Belle Vue (see Samuel Alberti's essay in this volume). In the early nineteenth century, most naturalists considered the ZSL's museum superior to the British Museum and the perfect place to lodge their specimens. As Charles Darwin wrote to a colleague upon his return from the *Beagle* voyage in 1836, "The Zoological Museum is nearly full & upward of a thousand specimens remain unmounted. I daresay the British Museum, would receive them, but I cannot feel, from all I hear, any great respect even for the present state of that establishment."[15] This remained the case until 1855, when the ZSL chose to close its museum and disperse its wealth of animal material. There were two reasons for this move. First, there was so much of it that their accommodation—by now on the west side of Leicester Square—was wholly inadequate. Second, and perhaps more important, the keeper of zoology, John Edward Gray, had done wonders at the British Museum, which had by this time come to be considered Europe's preeminent zoological repository. So the ZSL sold its most important specimens to the British Museum for £500. It is by this rather circuitous route that interesting animals dying at the zoo came to the BM(NH).[16] It is a tradition that has continued ever since, with the zoo's most celebrated inmates wending their way in death across the capital. There have also been many zoologists, such as Richard Owen, William Flower, Terence Morrison-Scott, Ronald Hedley, and Michael Dixon, who have led both the ZSL and the BM(NH), a crossover that has surely strengthened the bond between these two key institutions. Within a week of Chi-Chi's death, the museum had decided it would put her on display. This sat nicely with the recent drive for greater public engagement.[17]

When the mineralogist Frank Claringbull had assumed the director's chair in 1968, he took on the responsibility for giving the public face of the museum a much-needed makeover. Ultimately, he was responsible for a strengthening and wholesale reorganization of the Exhibition Section, which after 1973 would provide a strategic vision for the galleries and draw on the scientific expertise of more than a single department. This led to a stream of grand exhibitions from the late 1970s onward, including one entitled Human Biology (which thrilled the majority of the public and scandalized a small minority) and the immensely popular Dinosaurs and their Living Relatives, the sort of display the museum would not dare be without today. It is this same engagement ethos, which has become stronger with each passing year, that is responsible for the Thames Whale phenomenon (see Richard Sabin's essay in this volume).

But when Chi-Chi arrived in 1972, Claringbull was still busy paving the way for these reforms. He had pushed for a new Children's Centre, a waiting area near the main entrance, a bookshop, and a snack bar. And in the absence of a beefed-up Exhibition Section, he was personally involved in the creation and marketing of all exhibitions. This explains why, on 27 July, he issued a press release to reveal the museum's plan for Chi-Chi: "The skin will be mounted and put on display to the public as soon as possible but this process will take several months and the skin will not be available for inspection during this period. The skeleton will be added to the study collection and will be available only for research purposes."[18] At the bottom of the release, the museum offered up the curator of mammals, Gordon Corbet, as a source of "further information." Within days of the press release going out, he found himself fielding calls from the media asking if "special arrangements can be made to photograph her in various stages of preparation before going on display."[19]

The museum responded to the media interest and began to hatch a grander plan for Chi-Chi's display. By mid-September, it was agreed that they should aim to have the exhibit ready to show off at the next trustees meeting on 12 December and reveal it to the public the following day. At such meetings, the director asked the trustees to approve his policies and resolutions for the coming months, and it was always nice to have a spectacular new exhibit with which to bring them onside. Unveiling Chi-Chi in mid-December would also have the advantage of enticing schoolchildren to the museum once they'd broken up for Christmas.

With this in mind, the museum made a couple of crucial changes in how they were to handle Chi-Chi's remains. Her skeleton, which had formerly been destined for the research collection in the museum's subterranean vaults, now found itself part of an expanded public display that would parade her bones as well as her mounted skin. In addition, the museum appears to have responded directly to the media's requests to witness the taxidermy. Contrary to Claringbull's first Chi-Chi–related press release, his second (sent out in early October) announced that there was to be a photo call at the Modelmaking and Taxidermy Section based in Cricklewood the following week.[20]

The exhibition officer Michael Belcher kept his London Zoo counterparts informed of this development, and Tony Dale of the zoo sent two members of the press team out to Cricklewood to meet Roy Hale, the museum's taxidermist who had skinned and measured Chi-Chi's carcass during the post-mortem. "We shall all be interested to learn what press coverage you obtain following the photocall," he wrote to Belcher.[21] At the photo call, there was widespread interest in the process of taxidermy, largely because of Hale's enthusiasm. "A taxidermist is carpenter, metalworker, seamstress, sculptor and anatomist all in one," he told his audience.[22] Writing for the *Kensington News & Post,* Joanna Lyall made the taxidermist rather than the panda the focus of her story. And Georgina Wilson of the Australian Broadcasting Corporation, who interviewed Hale for a radio program on Chi-Chi, considered him "most helpful, and very good radio 'talent'."[23]

## THE TAXIDERMY

When it comes to taxidermy, it is notoriously hard to do justice to large mammals. In the nineteenth century, it was common practice to construct a wooden frame from which to hang the skin, then stuff it with straw or paper. But it would be very hard to capture the shape of an animal correctly and harder still to convey any impression of movement. Since the skin tightens as it dries, the stitching would inevitably begin to show. So poor were such preparations, in fact, that William Henry Flower, the director of the BM(NH) in the last decades of the nineteenth century, delivered the following damning criticism in his presidential address to the British Association for the Advancement of Science in 1889: "I cannot refrain from saying a word upon the sadly-neglected art of taxidermy, which continues to fill the cases of most of our museums with wretched and repulsive caricatures of mammals and birds, out of all natural proportions, shrunken here and bloated there, and in attitudes absolutely impossible for the creatures to have assumed while alive."[24] By the end of the century, however, the state of the taxidermic art had begun to improve. Based on precise measurements of the skinned carcass, the taxidermist would construct a skeletonlike "manikin" from a combination of wood and wire and then build

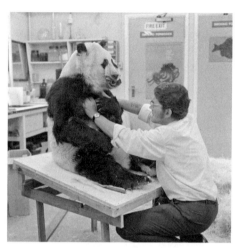

FIG. 1. Roy Hale, the senior taxidermist at the British Museum (Natural History), pulls Chi-Chi's skin over the manikin before her unveiling in the North Hall on 12 December 1972. (Natural History Museum Archives, PH/139/9; © The Natural History Museum, London)

up the muscles with "wood wool" (see fig. 1). The next step was to smear a thin layer of wet clay over the manikin. "The skin should then fit over the model like a glove," explained Hale, who had learned his trade working for Rowland Ward Ltd. of Piccadilly, then the most highly respected taxidermy enterprise in the world. It could then be pressed onto the wet clay and sculpted into the desired shape. For Chi-Chi's head, Hale would mold her skin to a fiberglass cast of her skull, taking great care to get her expression just right. "So many people have seen her in the zoo that they will soon say if they don't think she's lifelike when I've finished," he told his audience.[25]

Following the photo call, Belcher was pleased with his orchestration of the media, purring to Tony Dale at the zoo that "we got the press coverage we hoped for."[26] That is, with one notable exception. The *Daily Express* columnist Jean Rook—also known as the "First Lady of Fleet Street"—rarely missed an opportunity for satire. "My readers," she later wrote, "love me to sink my teeth and typewriter keys into some public figure they're dying to have a go at themselves, more especially if it's some sacred cow—or bull—who's never criticised by journalists."[27] On this occasion, Rook championed Chi-Chi, who "kept her virginity in an age when all around were losing theirs" and laid full square into the panda's new custodians: "Now that she's dead, dead and they'll never call her mother I think it's indecent of the Natural History Museum to exploit the remains when all that's left is a fibreglass model inside a giant fur coat," she wrote. "In life she may have chosen, in fact fought tooth and claw, to stay on the shelf. That's still no excuse for having her dusted."[28] Dale responded to Belcher's letter with this wry consolation: "I am sorry about Miss Rooke [*sic*]—if she appears again she could always end up in the polar bear pit!"[29]

## THE DISPLAY

Meanwhile, the staff back at the museum responsible for the display material that would make up the two display cases were working away to meet the December deadline. The responsibility for coordinating the effort fell to Exhibition Officer Belcher. On 20 September, he wrote to the keeper of zoology to kick-start a discussion over the precise contents of the display and to request a rough outline of any text and labels by 13 October. This would give everyone a couple of weeks to reflect on things before he would have to send the finalized text to the printers in early November. "If the dates can be kept, we will endeavour to have the display open for the Trustees Meeting on 12 December," wrote Belcher.[30]

There would be two display cases, one on each side of the archway leading north from the museum's North Hall. The scientists decided that the left-hand display case would house Chi-Chi's articulated skeleton and address the then unresolved question, "What, exactly, is a panda?" Belcher's team used birch blockboard to construct the display and some cognac-colored cloth to cover it. This was to provide the tasteful backdrop for the panda's skeleton and for some skulls and bones from red pandas, black bears, and raccoons. Giving the public a chance to look at the similarities and differences between these species might help explain how museum scientists went about their taxonomic business and might even communicate just how difficult it was to put the panda in its proper evolutionary place. Gordon Corbet, meanwhile, prepared the text and gave permission for real skulls to be used in the short term, though insisted that they "be replaced with casts when ready."[31]

In the right-hand display case, the Exhibition Section sought to create a "habitat diorama" for Chi-Chi's mounted skin (see fig. 2). With time of the essence, however, it was to be a struggle to make her surroundings realistic. They bought up some artificial fronds of leafy bamboo to create a bush to the left of Chi-Chi, some lengths of genuine bamboo cane that would be propped up to the right, and three bags of granulated peat to cover the floor beneath the panda's bottom. Alongside the scenery and the text, the Exhibition staff

FIG. 2. Chi-Chi on display in the Natural History Museum, 2010. (Author's photograph; © The Natural History Museum, London)

selected six photographs to illustrate Chi-Chi's life and fame. In one of them, taken during her brief stop-off at Frankfurt Zoo on the way to London in 1958, she is standing on her hind legs as if about to shake hands with *La Dolce Vita* star Marcello Mastroianni. An artist was commissioned to paint a mountainous backdrop and came up with the idea of incorporating the Chinese symbol for giant panda into it.

By mid-November, somewhat behind the schedule Belcher had outlined at the start, things were coming together. There was just enough time for Claringbull to be confident the exhibit would be unveiled in time for the trustees meeting. On 8 December, he issued his final Chi-Chi–related press release alerting journalists to another photo call on the morning of 12 December before the new display was opened to the public the following day.[32]

That might have been the end of Chi-Chi's journey, except that one last outing had already been arranged. In late November, the BBC had contacted the museum to find out if they could feature the all-new Chi-Chi on their flagship children's show. "*Blue Peter* is a reputable programme and is watched by some 9,000,000 children," wrote Belcher to the museum director, Frank Claringbull. But, he explained, "all their programmes are produced live and they are not interested in coming here to film." On top of that, they wanted

Chi-Chi to appear on the Monday show, the day before she was due to sweeten the trustees. The keeper of zoology had no objections to the giant panda making a final journey to West London, "providing it is at all times accompanied by a member of the museum staff at the expense of the BBC."[33]

Belcher now sought the go-ahead from Claringbull, who clearly gave his blessing because on Monday morning a BBC production assistant arrived at the museum in a van with a requisition note for "One stuffed Giant Panda." Peter Purves was the *Blue Peter* presenter who got the job of introducing Chi-Chi to the cameras: "Lots of people were rather sad when Chi-Chi died, so it's good to know that she's been so beautifully preserved and that she'll be on display at the Museum for ever and ever," he comforted his youthful audience.[34] After the show, Chi-Chi was escorted back to the museum and placed inside her display case to face the press, then the trustees, and finally the public.

Over the decades that have passed since then, the BBC has returned repeatedly to this zoological wonder. In 1992, the BBC broadcast *Chi-Chi the Panda,* a documentary that relived Chi-Chi's years at London Zoo.[35] In March 2009, a BBC Radio 4 documentary, *Panda Ambassador,* dipped into Chi-Chi's "life after death."[36] Then, in February 2010, museum staff opened up her case to allow a BBC TV crew to get up close for its six-part series *The Museum of Life.*[37] Celebrity pig farmer–turned–TV presenter Jimmy Doherty climbed into the case to carry out a "condition check" on the specimen. Chi-Chi's taxidermied skin remains in fine form.

### THE PUBLIC REACTION

It is surprising that in all this there was not more disquiet from the public or anxious nail-biting on the part of the museum's curators. Things could not have been more different a few years later when another famous London Zoo animal—Guy the gorilla—died from a heart attack during an operation to remove a rotten tooth. Within days of his death, word had leaked out that the BM(NH) was to give him a taxidermic overhaul, and the headline writers guffawed at the idea of "stuffing the Guy." The British public was incensed, and the zoo began to be inundated by a stream of vitriolic mail, all of which demanded a conciliatory and time-consuming personal response. Easily the most sensational was a letter from Viscount Anthony Chaplin, a past honorary secretary of the ZSL, who wrote to his successor to express how "utterly revolted" he was that Guy's remains were to be handed over to a taxidermist.

Chaplin would rather Guy were "buried or cremated and a memorial of some sort, however modest, placed in the Gardens of the Society." He went on: "Guy has been a friend of thousands since he was a child. I have known him since he was 2 or so. Would you, Sir, be complacent if somebody you had known since their childhood were submitted to a taxidermist to be put on show? Was Guy asked what his wishes were in this context? Did he leave a 'will'"? Chaplin signed off with an even stronger rant: "Are all future Hon. Secs, Presidents etc. of the Society to be stuffed and exhibited in a museum?"[38]

So why was it that Chi-Chi's taxidermy did not cause a stir while Guy going under the knife just six years later triggered such intense outrage? In large part, of course, it's simply due to the fact that one is a gorilla and the other a panda; we humans are just that much more touchy about the treatment of our close living relatives. In addition, the 1970s experienced a surge in interest in the great apes. The work of the primatologists Jane Goodall and Dian Fossey brought new knowledge of the intricacies of chimpanzee and gorilla societies, leading to the foundation of the Jane Goodall Institute and The Dian Fossey Gorilla Fund toward the end of the decade. It was also in 1979 that the BBC's Natural History Unit broadcast its most ambitious wildlife series to date—*Life on Earth*—which starred a young David Attenborough and featured his now-memorable encounter with Rwanda's mountain gorillas.

But this newfound wonder at the great apes was really just a reflection of bigger changes that were afoot as the environmental movement continued to gather momentum. The increasing frequency of natural history broadcasting during the 1970s—and in vibrant color—resulted in far greater public awareness of and sensitivity toward the natural world. Apart from Guy's brief appearance in the galleries of the BM(NH) in the 1980s, Chi-Chi was one of the last large mammals to be turned over to the museum's taxidermists for public display. By the end of the 1980s, stuffing animals for the public galleries had become so politically charged that the museum decided to shut down its Taxidermy Section for good, putting what little work they subsequently needed out to tender. Roy Hale, who had processed Chi-Chi more than fifteen years earlier, decided the time had come to retire.

CHI-CHI'S SPIRIT

While Chi-Chi's tissues and organs traveled the world and her skin and bones found a welcome home in the North Hall of the Natural History Museum, there is another less tangible component to this personality's afterlife: her spirit. Although Chi-Chi left no panda offspring, she turned out to be remarkably productive in other ways.

With Su-Lin causing such "panda-monium" at Brookfield Zoo in Chicago in the 1930s, other zoos were prepared to go to extraordinary lengths and to incur great costs to show off a panda of their own. When the Chinese Communist Party came to power in the aftermath of World War II, however, there was an end to the movement of pandas from East to West. Chi-Chi's fame reignited the West's desire for everything panda. Manufacturers quickly sensed the opportunity and began to turn out panda postcards, key rings, and soft toys. It was in 1964 that the hugely popular BBC TV children's program *The Sooty Show* added a panda, "Soo," to its cast of central puppets. Simultaneously, the vast and tittering media attention surrounding Chi-Chi's failed fling with An-An helped turn the giant panda from a zoological curiosity into a target for playful, sometimes wicked satire.

In October 1966, when zookeepers in Moscow introduced Chi-Chi to An-An, the headline writers had a field day. "Chi-Chi Is Playing Hard to Get," "Chi-Chi Gives An-An a Cuff," "Chi-Chi's Right Hook for the Suitor," they variously announced after the pandas' first encounter.[39] They had even more fun the next day when it was decided that the pandas should bed down in the same enclosure: "Two Pandas Spend Night Together"; "Pandas' Night of Promise"; "Strangers in the Night."[40] As panda relations turned from bad to worse, the headlines took on a gloomier tone: "Time Runs out for Chi-Chi"; "Chi-Chi Has Only Three Nights Left"; "From Russia—Without Love."[41] And then it was announced that Chi-Chi would be heading back to London: "Bride Who Never Was Flies Home"; "Return of the Virgin Panda"; "Chi-Chi, An-An Say Ta-Ta."[42]

The cartoonists, too, sharpened their pencils, contributing a spate of sketches throughout Chi-Chi's stay in Moscow and during An-An's return visit to London in 1968. Some of these toyed with cold-war espionage. Others fed off changes taking place with respect to human reproductive medicine and women's position in society. There were dozens of cartoons pegged to political developments, with a long list of politicians interacting with and variously cast as either Chi-Chi or An-An, including British prime minister Harold Wilson; his foreign secretary, George Brown; the leader of the opposition, Ted Heath; Rhodesian prime minister Ian Smith; and the Soviet premier Alexei Kosygin to name but a few. The satire on show in these cartoons offers a striking parallel to that exposed by the "Queen's Ass" some two centuries earlier (see

FIG. 3. In 2009, the Worldwide Fund for Nature decommissioned its panda collecting boxes, which—it could be argued—were modeled on Chi-Chi's image. Artist Jason Bruges used an array of these boxes for his interactive installation *Panda Eyes,* which was shortlisted for the Brit Insurance Designs of the Year 2010 at the Design Museum in London. (Photograph by Luke Hayes; *Brit Insurance Designs of the Year 2010,* Design Museum [www.designmuseum.org])

Christopher Plumb's essay). Finally, there was a whole other category of cartoons that simply made fun of the pandas. A perfect example appeared the day after An-An left London for Moscow in May 1969. The *Evening Standard's* JAK had Chi-Chi sitting alongside An-An's empty enclosure and imagined her thinking, "Gosh, I feel so sexy today!"[43] This explosion of farce surrounding and explicitly directed at the giant panda may help account for the origin of the popular myth that this is a species with an unhealthy dependency on bamboo, little or no interest in sex, and, as such, deserving extinction.

Finally, there is at least one other way in which Chi-Chi's spirit lives on. On 15 July 1961, Gerald Watterson, the secretary-general of International Union for Conservation of Nature and Natural Resources, went to stay with the environmentalist and artist Peter Scott at his home in Slimbridge on the Bristol Channel. There Watterson showed Scott some sketches of a panda—almost certainly based on Chi-Chi—which inspired Scott to create the first logo for the World Wildlife Fund (WWF), which became a legal entity later that year (fig. 3).[44] Through this brand and the funds that WWF has ploughed into the conservation of giant pandas, Chi-Chi has had and continues to have a profound impact on the fate of this species today. More than that, it's not unreasonable to argue that this singular panda, as the face of global conservation, has had a hand in every campaign the WWF has ever run, from species-based projects like Operation Tiger to habitats, ecosystems, and the globe (see Hannah Paddon's essay in this volume for a detailed consideration of mascotism).

For most people alive today, Chi-Chi's life story, if known at all, is little more than a peculiar footnote in history. Thinking about what happened to this zoological specimen after her death, however, illustrates perfectly how much more there is to be gained from thinking about the afterlives of animals. Chi-Chi's physical remains, which led to the publication of several scientific papers in the 1970s, have made a small but significant contribution to the global research effort to understand the giant panda. The BM(NH)'s treatment of her skin and bones offers an intriguing glimpse into museum practice in the 1970s and an opportunity to reflect on the communication of science then and now. Looking beyond the immediate details of the Chi-Chi/An-An affair suggests that this celebrity couple may have had a lasting influence on the public perception of the giant panda, particularly in Britain, where this species continues to cut a very comedic image. Finally, through the WWF's panda logo, Chi-Chi became the face of the world's largest nongovernmental conservation organization. It is her image that has inspired millions to reflect on humankind's position among other species and our profound impact upon them.

NOTES

1. Ian F. Keymer, "Report of the Pathologist, 1971 and 1972: The Scientific Report of the Zoological Society of London, 1971–1973," *Journal of the Zoological Society of London* 173 (1974): 45. For a more detailed account of Chi-Chi's (premortem) life, see Henry Nicholls, *The Way of the Panda: The Curious History of China's Political Animal* (London: Profile, 2010).

2. *Nationwide,* BBC1, April 1972.

3. Dwight D. Davis, "The Giant Panda: A Morphological Study of Evolutionary Mechanisms," *Fieldiana Zoology Memoirs* 3 (1964): 1–337.

4. Keymer, "Report of the Pathologist."

5. "Chi-Chi the Panda," *Arena,* BBC2, 20 March 1992.

6. "British Panda Chi-Chi Dies," *Star-News,* 23 July 1972.

7. Ian F. Keymer, "'Chi-Chi': Pathology," *Transactions of the Zoological Society of London* 33 (1976): 103–18.

8. Herbert J. A. Dartnall, "Visual Pigment of the Giant Panda *Ailuropoda melanoleuca,*" *Nature* 244, no. 5410 (1973): 47–49.

9. Angela S. Kelling et al., "Color Vision in the Giant Panda (*Ailuropoda melanoleuca*)," *Learning and Behavior: A Psychonomic Society Publication* 34, no. 2 (2006): 154–61.

10. Norman H. Ashton, "'Chi-Chi': The Eye," *Transactions of the Zoological Society of London* 33 (1976): 127–31.

11. Leonard G. Goodwin, ed., "'Chi-Chi': The Giant Panda *Ailuropoda melanoleuca* at the London Zoo 1958–1972: A Scientific Study," *Transactions of the Zoological Society of London* 33 (1976): 77–171.

12. Davis, "The Giant Panda," 16.

13. Stephen J. Gould, "The Panda's Peculiar Thumb," *Natural History,* November 1978, 106.

14. Vincent M. Sarich, "'Chi-Chi': Transferrin," *Transactions of the Zoological Society of London* 33 (1976): 165–71.

15. Charles R. Darwin to John S. Henslow, 30 October 1836, Darwin Correspondence Project Database letter 317, www.darwinproject.ac.uk/entry-317.

16. Henry Scherren, *The Zoological Society of London: A Sketch of Its Foundation and Development and the Story of Its Farm, Museum, Gardens, Menagerie and Library* (Cassell, 1905); Sofia Åkerberg, *Knowledge and Pleasure at Regent's Park: The Gardens of the Zoological Society of London during the Nineteenth Century* (Umea: Universitets Tryckeri, 2001).

17. William T. Stearn, *The Natural History Museum at South Kensington: A History of the British Museum (Natural History) 1753–1980* (London: Heinemann, 1981).

18. G. Frank Claringbull, "Chi-Chi at the Natural History Museum," 27 July 1972, Natural History Museum Archives, DF 700/106.

19. A. Clarke to Michael Belcher, 7 August 1972, NHM Archives, DF 700/106.

20. G. Frank Claringbull, "Chi-Chi at the Natural History Museum," 5 October 1972, NHM Archives, DF 700/106.

21. J. A. Dale to Michael Belcher, 9 October 1972, NHM Archives, DF 700/106.

22. Joanna Lyall, *Kensington News & Post,* 12 October 1972.

23. Georgina Wilson to Michael Belcher, 4 November 1972, NHM Archives, DF 700/106.

24. William Henry Flower, *Essays on Museums and Other Subjects Connected with Natural History* (London: Macmillan, 1898), 17.

25. Joanna Lyall to Michael Belcher, 12 October 1972, NHM Archives, DF 700/106.

26. Michael Belcher to J. A. Dale, 13 October 1972, NHM Archives, DF 700/106.

27. Ann Godden, "Jean Rook: The First Lady of Fleet Street," 1991, www.hullwebs .co.uk/content/l-20c/people/Jean%20Rook.pdf.

28. Jean Rook, *Daily Express*, 12 October 1972.

29. J. A. Dale to Michael Belcher, 17 October 1972, NHM Archives, DF 700/106.

30. Michael Belcher to J. Gordon Sheals, 20 September 1972, NHM Archives, DF 700/106.

31. D. Gosling to Michael Belcher, 7 November 1972, NHM Archives, DF 700/106.

32. G. Frank Claringbull, "Chi-Chi at the Natural History Museum," 8 December 1972, NHM Archives, DF 700/106.

33. Michael Belcher to G. Frank Claringbull, 23 November 1972, NHM Archives, DF 700/106.

34. "Blue Peter," BBC1, 11 December 1972.

35. "Chi-Chi the Panda," *Arena*.

36. "Chi-Chi: Panda Ambassador," BBC Radio 4, 11 March 2009.

37. "Museum of Life," BBC2, www.bbc.co.uk/programmes/b00rp1w0.

38. Anthony Chaplin to Ronald H. Hedley, 6 November 1978, NHM Archives, PH/219.

39. "Chi-Chi Is Playing Hard to Get," *Oldham Evening Chronicle*, 7 October 1966; "Chi-Chi Gives An-An a Cuff," *Swindon Advertiser*, 7 October 1966; "Chi-Chi's Right Hook for the Suitor," *Newcastle Evening Chronicle*, 7 October 1966.

40. "Two Pandas Spend Night Together," *Gloucester Echo*, 8 October 1966; "Pandas' Night of Promise," *Shields Gazette*, 8 October 1966; "Strangers in the Night," *Birmingham Mail*, 8 October 1966.

41. "Time Runs out for Chi-Chi," *Hull Daily Mail*, 11 October 1966; "Chi-Chi Has Only Three Nights Left," *Citizen*, 11 October 1966; "From Russia—Without Love," *Bath and Wiltshire Chronicle*, 11 October 1966.

42. "Bride Who Never Was Flies Home," *Press and Journal*, 18 October 1966; "Return of the Virgin Panda," *Morning Star*, 18 October 1966; "Chi-Chi, An-An, Say Ta-Ta," *Staffordshire Evening Sentinel*, 17 October 1966.

43. "Gosh, I Feel So Sexy Today!" *Evening Standard*, 22 May 1969, 15463, British Cartoon Archive.

44. The first mention of Gerald Watterson's role in the creation of the World Wildlife Fund's panda logo appears to be in a brochure produced for the organization's twentieth anniversary: *World Wildlife Fund Twentieth Anniversary Review*, Max Nicholson Archive, Linnean Society of London, EMN 4/19/1,2.

RICHARD C. SABIN

# The Thames Whale

## The Difficult Birth of a Celebrity Specimen

On the morning of 19 January 2006, the Natural History Museum's Whale Strandings hotline received a telephone call from Thames Coastguard. The caller gave details of an earlier sighting of several whales at the mouth of the Thames estuary. Later the same day, staff at the Thames Barrier reported that one or two animals had been seen some fifteen miles upriver passing through the barrier. Over the course of the following two days, a sequence of events took place which captivated many of the occupants of one of the biggest cities in the world. As news spread across the globe, international media besieged the banks of the River Thames in central London. Though unaware of it at the time, the thousands of ordinary people who flocked to the Thames were staking their claim to an "I was there" moment, as a new social memory was shaped and nurtured, eventually leading to the birth of the Thames Whale. This essay is an account of my professional and personal involvement in the story of the Thames Whale. It examines the processes by which a wild animal from the open ocean became both celebrity and scientific research specimen.

As curator of marine mammals in the Department of Zoology at the Natural History Museum, London, I deal with specimens that have been a living part of a vibrant and dynamic ecosystem. I seek to enhance the scientific value of these specimens by making them accessible for use in global taxonomic and conservation research. Though this is one of the main objectives of my work, it does not preclude developing an interest in how such scientific collections came to be: few specimens are without an interesting history. Each represents the efforts of individuals or expeditions to explore, collect, catalogue, organize, and understand the natural world. As a whole, for more than two centuries, such collections of specimens have formed a valuable and vital scientific resource. They continue to be crucial to the activities of researchers, forming a constantly developing model of the natural world, rather than simply being artifacts of past activities. This is the framework into which the Thames Whale was introduced.

## THE WILD COMES TO LONDON

In 1913, scientists at the British Museum (Natural History) in South Kensington asked the British government for access to the carcasses of those whales, dolphins, and porpoises—collectively known as the scientific order Cetacea—which were found washed ashore around the coasts of the British Isles. Negotiations began with representatives of the Crown, as stranded cetaceans were classified as "Fishes Royal" under a prerogative enacted in 1324 during the reign of Edward II.[1] Agreement was eventually reached between the museum and all other parties that museum scientists could have access to the carcasses of stranded cetaceans for the purposes of scientific investigation. The museum was to collate data and report annually to the government, procedures that continue to the present day.

So it was that on 19 January 2006, a telephone call was made to the Natural History Museum by Thames Coastguard, alerting us to the presence of large cetaceans in the Thames estuary. January 2006 had been a relatively quiet month in terms of the number of stranded cetaceans reported to the museum. The initial report of several large whales sighted at the mouth of the Thames estuary was alarming, and as coordinator of the museum's Cetacean Strandings Monitoring Project, I remember hoping the report would come to nothing. Later that same day, a subsequent report was made stating that one or two whales had been observed passing through the Thames Barrier, heading upriver. I was on my way home when I picked up this second call, still hoping that whatever was in the river would follow the tide back out to sea.

Early on the morning of 20 January, the museum's press office received a telephone call requesting my presence for an interview. A television news crew had set up base at Chelsea Bridge, approximately ten miles from the Thames

Barrier and only three miles from the center of London. They had footage of a whale they had filmed a few minutes earlier, after receiving information that a rail commuter had seen a large animal in the Thames from the window of his carriage. The television crew said the commuter was concerned that he had suffered a hallucination. For me, the news was a shock. I traveled to Chelsea Bridge with a museum press officer, hoping that whatever we saw would be small and easy to handle. How wrong I was. The television crew excitedly explained they had footage of an animal they had been told by their studio was a pilot whale. I sat in the back of their outside broadcast van with the crew crowded around me and watched the video. I stared at the small screen; the gray surface of the Thames was broken by an even grayer, bulbous head with a characteristically long snout. The animal spouted, presenting a long expanse of back, a tiny dorsal fin, and two enormous tail flukes before diving below the surface. I remember sitting transfixed and hearing myself say, "Oh hell, it's a northern bottlenose whale. We have a very big problem."

The next half hour was a flurry of telephone calls; informing our veterinary colleagues at the Zoological Society of London of the animal's identification so they could coordinate with rescue teams and prepare their welfare response accordingly; contacting the museum's database manager and requesting a breakdown of all cetacean species we'd recorded in the Thames over the past one hundred years; compiling a museum press release to go out to the media, and, finally, calling my wife to let her know that I might be late home. The events of the six or seven hours that followed were bizarre, almost dream-like. Within an hour of my arrival, other television crews appeared and set up their equipment. As I completed one interview, reporters from another station would appear and drag me away. They all asked the same questions: What is it, why is it here, and will it live? Museum data confirmed that the northern bottlenose whale, scientific name *Hyperoodon ampullatus,* had never before been recorded in the Thames.

As news of the whale began to filter out, people started to gather along the Thames from Chelsea to Battersea Bridge, a distance of around one and a half miles. Office windows were thrown open by workers eager to catch a glimpse. As I completed a radio interview, I turned in time to catch my first view of the whale. There were shouts and screams from the media and the public. A wave of people ran to the embankment wall, and cameras started to flash. I stood for a few minutes watching this beautiful animal sliding through the waters of the Thames, but my overwhelming feeling was one of impending dread. A short time later, I heard the first of the helicopters arrive. The pace of events suddenly changed. The whale was now a big news story, and resources did not seem to be in short supply. Small powerboats appeared in the river as the whale headed toward Battersea Bridge. Some boats were crewed by response teams from British Divers Marine Life Rescue, the organization that would eventually coordinate the rescue attempt; other boats were crewed by police, the Port of London Authority, or had been hired by journalists. I ran toward Battersea Bridge with a crowd of others to follow the whale; van drivers sounded their horns and shouted at the crowds; cars were abandoned in the road as drivers rushed to see what was going on. How must the scene have looked to those driving by who had not heard the news? Somewhere off in the distance, I heard someone shout, "Free Willy!" People were beginning to make their connections with the event, to have their moments and build their narratives.

At Battersea Bridge, I saw the whale lodged between large, moored vessels. The whale was thrashing its flukes and raising and lowering its head in an attempt to get free. Blood was in the water, and as the whale made its way back to midchannel, I saw a gash on the end of its snout. This caused distress to many of those watching, and several people were shouting for someone to do something. Being familiar with the behavior of the species, I was concerned that other whales downriver might be attracted to distress calls from this animal. The whale was trying to swim against the current. The tide had turned, and the current in the river was very strong. Though the whale was swimming hard, it made little progress. It then disappeared below the surface.

By midafternoon, the media were camped along the banks of the Thames between Battersea and Chelsea bridges. The information I had earlier authorized as a museum press release kept being reported back to me, but had begun to subtly change. The northern bottlenose whale became "the northern blue-nosed whale"; the first record of this species in the Thames became "the first whale ever seen in the Thames." More helicopters had appeared, and the authorities had difficulties controlling river traffic. At around 4:00 PM, as the light began to fade, the whale was seen swimming downriver toward Westminster Bridge. Earlier in the day, it had been photographed as it swam

FIG. 1. The Thames Whale in front of the Houses of Parliament, London, 2006. (Image by Liz Sandeman [the first person to photograph the whale]; © Marine Connection)

in front of the Houses of Parliament (fig. 1). That photo was to become one of the iconic images of the event, leading the media to refer to the whale as an ambassador from the sea.[2]

Around 6:00 PM, with the light fully gone, word was passed around that the whale could not be located and that efforts would be resumed the next day. I had completed more than twenty interviews with media from as far afield as Russia, Canada, and Brazil. My last interview of the day was with BBC Radio. It was during this interview I was asked if I had a thought of a name for the whale. I said that I had not, but the presenter said many listeners had already called in with suggestions, including (inevitably) Willy, Whaley, Wally, and Roody, the last apparently taken from the generic name *Hyperoodon*. I was quietly impressed by this. All the names sounded rather masculine, and for the first time I was aware that there had been no consideration by the media that the animal might, in fact, be female.

## THE BIRTH OF SPECIMEN SW.2006/40

At 5:30 AM on Saturday, 21 January, I received a telephone call from the Natural History Museum's press office. They told me the whale had been spotted again near Chelsea Bridge and that the rescue team was planning to evaluate the situation as soon as it got light. Exhausted from the day before, I accepted the press office's offer to divert screened calls to my home. They had a list of interview requests plus calls from Sky News and the BBC asking if I would be available for comment on the unfolding events of the day. I agreed. This was to herald a day spent sitting down, drinking tea, and eating toast.

By around 7:30 AM, Sky News had already begun reviewing the events of the previous day. They had an aerial view from their helicopter which showed the walkways and bridges along the Thames filling up with people. Eventually, the sun came up, and once again the circus came to town. The first questions of the day were based on speculation that had begun to circulate the day before that some kind of human activity had caused the whale to stray off course. Reports were claiming that noise pollution may have scared the animal into the comparatively shallow North Sea, where it then traveled south until able to head west into the Thames estuary. Its efforts to swim upriver against the current had been interpreted as an inbuilt migratory mechanism to head west out into open water. There was much speculation, so in the absence of the facts and in the face of leading questioning, I stuck to what we knew about the whale: its species and what our archival data told us.

At around 9:30 AM, the whale was sighted in the river near Albert Bridge. Its movements were less powerful than the previous day, and word from the rescue team suggested the animal was exhausted. News then came through that inflatable pontoons had been drafted in from across the country to aid in the rescue. Port of London Authority had offered the use of a barge with a crane attached to lift the animal from the water. With this news, there seemed to be a growing realization that what had been an exciting spectacle the day before was now turning into a grave situation.

By midafternoon, the rescue team had secured the inflatable pontoons on either side of the whale and had lifted it aboard the barge, lowering it onto a base of inflatable rafts. The barge sped downriver, trying to stay ahead of the rising tide, which threatened to prevent navigation under low bridges. On television, impressive simulated graphics showed how the animal had been lifted and how it would eventually be released into the open sea. Commentators variously applauded and criticized the rescue operation. Time continued to pass with occasional reports coming from the barge. Suddenly, around 6:45 PM, though the best available care and attention had been given, the animal died. The lights on the barge were turned off, and the news was passed to a stunned audience. Within minutes, questions were asked about the fate of the whale's carcass. I made it clear that a full postmortem examination would be carried out by a veterinary team from the Zoological Society of London as part of the United Kingdom Cetacean Strandings Investigation Programme. Questions about the final fate of the carcass persisted, however, and I became aware that my response might not be to the liking of everyone. I was asked on several occasions if the Natural History Museum would take the body of the whale. My response was a combination of realism and regret. I said that we would work with our veterinary colleagues and examine the situation after the postmortem had been completed. I added that small tissue samples would be taken for DNA analysis, but the necessary resources were not available to us to prepare the entire specimen for our scientific research collections. This was my first public use of the word "specimen" in relation to the whale. It had crossed the line from living animal to inanimate thing. "Why don't you stuff the whale or pickle it?" I was asked. Though I used simple language to explain the principles of natural history research collections, I was continually pressed on the stuffing and pickling issue. I realized I was battling against deep-rooted stereotypes, and images of natural history collections which were embedded in the media and public mind-set.

At the end of the day's activities, I issued the official reference number for the stranding event: SW.2006/40. The "SW" number (denoting "stranded whale") is a unique reference given to each reported cetacean stranding, whatever the circumstances of that event. Like the naming of a newborn child, if anything was to mark the transition from wild, living creature to scientific study specimen, it was this; the true birth of the animal-as-object, and the completion of the first stage of its incorporation into the museum's research collections.

## SAVE THE WHALE

At 10:00 AM on 22 January, I was called by the museum's press office with another list of interview requests. This time, they demanded to know what would happen to the body of the Thames Whale. The idea of it being dumped in a landfill site was abhorrent to most. Though I had pointed out that this was the fate of dozens of carcasses of dead whales and dolphins washed ashore each year, the Thames Whale had already been elevated to some kind of special celebrity status.

The next telephone call was from the *Sun*, a national British tabloid. A polite young journalist asked me why we could not preserve the entire carcass of the whale for exhibition. After outlining the impracticalities of "pickling" the carcass of a three-ton whale measuring nineteen feet (five meters), I convinced him that the skeleton would be of most value to us for scientific research. The journalist asked me how much it would cost to preserve the skeleton. I said the work could run to a few thousand pounds. The journalist said he would talk to his editor and get back to me. An hour later, he called back and said that his team had come up with the idea of a "Save Wally for the Nation" campaign, where readers would be encouraged to donate money to be passed on to the museum. I said nothing about the name "Wally," but stated that the decision was not mine to make. I passed on details to colleagues at the museum, and later that day, I was informed that, after much careful debate, the offer had been tentatively accepted. There were, however, some conditions, chief among these being that the first photographs of the prepared skeleton should appear in the *Sun*.

With the prospect of funding, I contacted the veterinary team who were carrying out the postmortem at a closed location in Gravesend. The Port of London Authority had permitted the off-loading of the whale's carcass at one of their wharves, screened from public view. Agreement was reached that the carcass would remain at this location until I could field a team to carry out dissection to retrieve the skeleton of the animal.

Monday, 23 January was a day of bringing men, women, and sharp implements together. I put together a team of seven people (including myself) and arranged for us all to meet at dawn on Tuesday, 24 January at the wharf in Gravesend. All had previous experience of field collection and preparation so knew a little of what to expect. I set off for Gravesend in the late afternoon and found accommodation at a pub less than a mile from the wharf. After settling in to my room, I went down for dinner. Since I was not a local, the bar fell silent as I entered. Within minutes, I had been asked where I was from, what I was doing, and would I like a drink? Mention of the Thames Whale turned me into a social magnet, and several people wanted to tell me their personal accounts of the whale's struggle. They were desperate to learn of anything new. I felt like an unknown sibling of a Hollywood star, dining out in reflected glory.

At sunrise on Tuesday, 24 January, we arrived at the wharf and presented ourselves to the Port of London Authority staff. They were hugely helpful, genuinely interested, and very accommodating. Large rubble sacks had been provided into which we would decant unwanted soft tissues, for eventual disposal at landfill. A static crane with a huge jib stood nearby, and we were told that when we needed to reposition the carcass they could assist. I was overwhelmed at the help we were receiving.

The day dawned cloudless, crisp, and very cold, with the only source of heat coming from the gently decomposing tissues of the whale's carcass. There was a general sense of detachment within the team; we had spent about a half hour at the beginning of the day discussing the objectives of the exercise. There followed a few minutes of involuntary silence before work commenced as individuals walked around the carcass, which had been left partially eviscerated by the postmortem of the previous day. The process of defleshing the carcass was straightforward enough; each person was allocated a portion of the carcass on which to work. Blubber and muscle tissue separated with ease as we worked. Particular care was taken to locate and remove the finer rudimentary pelvic bones and those of the hyoid region of the throat, which are often missing from many museum study specimens. The tendons which ran the length of the spinal column to the massive tail flukes were as thick as telegraph cables, and quickly blunted our blades. We worked without breaks.

By sunset, we were done. The bones of the Thames Whale had been removed, logged, wrapped, loaded into a van, and transported to frozen storage. I had dealt with a small number of journalists who tried to get in to the wharf as we worked, eager to photograph the initial stages of specimen preparation. I'd approached them and asked for their cooperation, as we felt it inappropriate that graphic scenes of defleshing should appear in the press.

So there we were: seven exhausted people, covered in an acceptable amount of blood (not our own) and all feeling extremely satisfied. There was a great sense of accomplishment; we had completed the second stage of the transformation of the whale into a museum study specimen.

## LIFE AS A RESEARCH SPECIMEN

On 25 January 2006, scientists at the Zoological Society of London released their initial postmortem findings for the Thames Whale. Their examination showed that the animal was a juvenile/subadult female, probably fewer than eleven years old and measuring 19 feet 3 inches (5.85 meters) in length. She had died as a result of several factors, including severe dehydration, some muscle damage, and reduced kidney function.[3]

In February, we began to plan for a journey north to Scotland. Colleagues at the Royal Museum in Edinburgh had kindly offered the rental of their workshop for preparation of the Thames Whale's skeleton. Funds being available, we gratefully accepted. Donations had also come from other sources; a Dorset school sent the Natural History Museum a check for £5 after having a class collection; a south London poetry and psychics group raised £12 after holding an evening of readings (in both senses of the word) at a local pub — the check was accompanied by a hand-drawn astrological chart, drafted especially for the whale.

On the morning of 17 March 2006, two of my colleagues and I loaded the frozen, roughly defleshed bones of the Thames Whale into a van and began driving north. The landscape turned snowy as we reached the Scottish border, and as we entered Edinburgh, we saw people making their way to bars and pubs to celebrate St. Patrick's Day. In the late afternoon, we off-loaded the skeleton of the whale and left one of my colleagues behind to spend the next week cleaning the bones. This was achieved using a popular brand of washing powder containing enzymes suitable for the controlled removal of soft tissues (fig. 2). The bones returned to the Natural History Museum store by the end of the month, where they were laid out on absorbent materials to collect the natural oil that oozed from them.

Interest in the specimen remained high. Letters arrived from schools around the United Kingdom, and requests from researchers for access to tissues for scientific investigation. In early June, the museum set up a photo call with the *Sun;* I duly articulated the bones of the whale, and the first public photographs of the specimen were taken.

Later in the summer, the museum was approached by documentary filmmakers who wanted to make a program about cetacean strandings, focusing on the Thames Whale event. Agreement was reached, and plans for filming

FIG. 2. Cleaning the bones with a popular brand of washing powder. (© The Natural History Museum, London)

were outlined. I was asked to once again articulate the skeleton of the whale for the camera and was interviewed on several aspects of strandings and the museum's collections (fig. 3). This was exactly what I had hoped for; an opportunity to highlight the importance of our research collections as a scientific and educational resource. The subsequent documentary was entitled *The Whale That Swam to London,* and was broadcast on 21 December 2006.[4] The documentary included contributions from all parties involved with the Thames Whale event and examined the whale's death in a global context.

In October 2006, the museum was contacted by the deputy editor of the *Guardian,* a national British broadsheet newspaper. They planned to mark the first anniversary of the Thames Whale's stranding with an exhibition of the skeleton and a series of public events. Their initial contact was to see if we were willing to loan the skeleton for display at the Newsroom, an exhibition space owned by the *Guardian,* just a few miles from the museum. My first thought was for the security of the specimen. I had a feeling that Thames Whale memorabilia would still be highly desirable, remembering the online

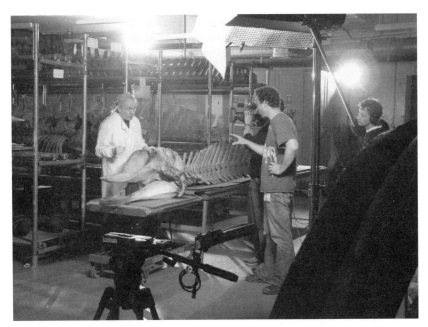

FIG. 3. The author filming *The Whale That Swam to London,* 2006. (© The Natural History Museum, London)

bidding war for the watering can used by rescuers to keep the whale hydrated as it was lifted from the Thames.[5] As a result, I stated that we would need a secure display case. After obtaining a cost estimate for the case, the *Guardian* agreed to pay for the construction. With the agreements in place, we began to plan the first public exhibition of the Thames Whale.

### LYING IN STATE: A CHANCE TO MOURN?

From 22 to 27 January 2007, the skeleton of the Thames Whale was on display in the Newsroom visitor center at the offices of the *Guardian* in London. Over the course of six days, approximately 2,500 people viewed the remains. Media interest was once again ignited, and though photography was prohibited, videos appeared on public blogs and file-sharing websites. Some were creatively set to music, while others were supported by comments from their viewers.

A series of public talks were held both at the Natural History Museum and the Newsroom. During one public event at the museum, a small boy sitting in the front row of the lecture theater grabbed the microphone, looked me in the eye, and shouted, "Why did you kill the whale?" I was speechless. However, the question was deftly handled by the event host, and I was free to go, but I was left wondering about the question and the way the boy had asked it. Had he condensed the events of that fateful weekend down to a single accusation? Through my association with the specimen, did he feel I was to blame for the death of the animal? There seemed to be an underlying sense of dismay from the audience that we had dissected the carcass and had only preserved the skeleton. The defleshing had somehow in their eyes reduced the value of the specimen as a representation of the Thames Whale. Surely, as a museum, was it not our responsibility to preserve the animal whole and display it for all to see in some approximation of life? If only it were that simple. Curators of vertebrate research collections have a series of choices to make when a specimen is offered to them, usually influenced by the species, condition, and size of the carcass. The hide may be removed and dry-preserved as a study skin; the soft tissues may be removed, subsampled, and placed into frozen storage for DNA analysis; the bones may be cleaned and preserved as a disarticulated study skeleton; the whole body may be placed into preservative fluid. Few museums preserve in fluid the whole bodies of very large mammals, as they are difficult to access, maintain, and use for detailed anatomical study. Larger mammals are usually represented either as parts or as young/fetal specimens.

Though the skins of all terrestrial and some marine mammals can be preserved using taxidermy, the same cannot be said for cetaceans. Early attempts by the Natural History Museum to do this met with failure. In 1922, Charles Tate Regan, keeper of zoology, explained in a museum guidebook that the proper exhibition of the Cetacea in a public museum was a matter of very great difficulty, not only because of the amount of space required by the larger species, but likewise owing to the difficulty of mounting the skins.[6] Initially, the mounted skins of whales and dolphins prepared in the late nineteenth century looked relatively representative, but over time they began to exude oil, discolor, and produce an unpleasant odor. By the 1930s, the museum had moved primarily to the use of skeletons and models in its whale displays, culminating with the completion of a life-size model of a blue whale in 1938. More than seventy years later, the blue whale model remains one of the most iconic and memorable specimens in the museum's galleries, acting as a point of reference for the recollections of many who have previously visited.

My own personal preference is for skeletal material. I see the vertebrate skeleton as the immutable framework of an organism which can provide so much information to scientists. There is honesty in skeletal material that I fail to see in taxidermy, which for me is the artist's representation of an animal (as are other models). Though the same may be said for articulated skeletons, which are equally an artist's representation and may show bones orientated entirely incorrectly, the individual bones have changed little since the animal's death. Herein lies the distinction between specimens prepared for exhibition and those for research purposes; taxidermy and articulated skeletons generally have a much lower research value. This was the message I was trying to convey to the press and public; that the value in the Thames Whale's remains lay in research, though we did not rule out displaying its skeleton at a later date.

CONCLUSION

In almost five years (at time of writing) since the public display of the Thames Whale skeleton, there has been little to disturb the animal's rest. It has become an important and integral part of the National Collection of Cetacea; it has been examined by a variety of individuals who intend to incorporate the data into their work; it has continued to leak oil from its bones as they sit in the case built to contain them.

Those from outside the museum who now see the skeleton in storage need little prompting to tell their personal "whale weekend" story, or to voice their opinions of how things could have been done differently. There is little doubt in my mind that people have invested varying degrees of emotion in the whale and what it represents to them, be it emblematic of the state of the world's oceans, a struggle to survive by a creature trapped in the middle of a city, or a reminder that there are still truly wild and awe-inspiring things in relatively close proximity to us.

The Natural History Museum continues to receive occasional letters and e-mails from the public and inquiries from the press asking about the status of the specimen. They want to know what more we have learned of the whale's life and death, assuming that the specimen was saved only for the purposes of researching these questions. In reality, the whale's skeleton will continue to generate scientific data for decades if not centuries to come; and in 2011 it was displayed once again, this time at the Natural History Museum at Tring.

The presence of the whale in the museum gives us the opportunity to communicate with a wide range of audiences. I continue to talk to groups of students about the diversity of cetacean species found in the waters around Britain, hoping that some will be inspired to go on to study marine biology or curate museum collections. Artists have sketched and photographed the bones of the whale, focusing on their aesthetic qualities rather than their scientific significance. Politicians occasionally make reference to the Thames Whale, as, for example, in an early day motion in the House of Commons in January 2007: "That this House remembers the anniversary of the Thames Whale; recognises the unprecedented public interest and concern for the plight of this and other whales generated by her visit."[7] In 2010, more than four years from the first appearance of a northern bottlenose whale in the River Thames, a Google search using the term "Thames Whale January 2006" brought up more than forty thousand hits. In that four-year period, I met almost as many people who think the whole thing was overblown and a complete waste of time as think it was a spectacular though fated event. There is little doubt that proximity was a factor; the sudden appearance of any large, wild animal in the center of any of the world's capital cities would be newsworthy. The explosion of mobile technology in the first decade of this century was doubtless a factor in the rapid spread of news about the whale, and in the huge public interest. By comparison, on 31 August 2006, two northern bottlenose whales were found stranded alive on a beach at Skegness on the English coast. The same veterinary experts and many of the leading rescuers in attendance during the Thames event went to Skegness to assist. Here, too, the animals died, but there was very little press or public interest.

The nature of celebrity can be ephemeral. It is affected by fashion and degraded by the ravages of time. The Thames Whale has become a celebrity specimen, joining the ranks of Chi-Chi the panda and Guy the gorilla; like the whale, these specimens have detailed, multilayered, and important narratives attached to them. Though time will pass and memories fade, their incorporation into the ranks of specimens at the Natural History Museum will ensure their stories are not forgotten.

NOTES

1. Francis C. Fraser, "Fishes Royal: The Importance of Dolphins," in *Functional Anatomy of Marine Mammals,* 3 vols., ed. Richard J. Harrison and Stephen Ridway (Academic Press, 1977), 3:1–41.

2. See BBC News Online, "Thames Whale Could Help Others," 22 January 2006. http://news.bbc.co.uk/1/hi/england/london/4637918.stm.

3. See Zoological Society of London, "ZSL Announces Thames Whale Post-Mortem Results," 25 January 2006, www.zsl.org/zsl-london-zoo/news/zsl-announces-thames-whale-post-mortem-results,233,NS.html; UK Cetacean Strandings Investigation Programme, "Northern Bottlenose Whale (*Hyperoodon ampullatus*)," in *Annual Report to Defra for the period 1st January to 31st December 2006 (Contract Number CRO346),* ed. Rob Deaville and Paul Jepson (London: Zoological Society of London, 2006), 16–18.

4. Firefly Productions, *The Whale That Swam to London,* broadcast 21 December 2006 on Channel Four.

5. Bidding reached £6,000 at one point (see Rosie Murray-West, "Frantic Bidding for Watering Can Used on Thames Whale," *Telegraph,* 24 January 2006).

6. Charles Tate Regan, preface to *Guide to the Whales, Dolphins and Porpoises (Order Cetacea) Exhibited in the Department of Zoology at the British Museum (Natural History), Cromwell Road, London, SW7,* 2nd ed., ed. Sidney F. Harmer (London: British Museum [Natural History], 1922), 3.

7. See Anne Main, "Thames Whale Anniversary and International Whaling Commission," 22 January 2007, Early Day Motion no. 694, EDM Database, www.publications.parliament.uk/pa/cm200607/cmwib/wb070127/edms.htm.

GARRY MARVIN

# Enlivened through Memory

## Hunters and Hunting Trophies

All of the animals explored in this volume have had afterlives that are longer, and more complex, than those of their species counterparts which died naturally, decomposed, or were eaten by other animals, or which were killed by humans and maybe eaten by them. Some of the animals discussed here, perhaps captured in the wild, did, once, have a natural life and were, once, wildlife. All of them, though (with the exception of the hen harrier), had cultural lives (albeit a short one in the case of the Thames Whale); lives lived in the presence of or with humans. They all had public lives which are now continued as afterlives in public, civic spaces, and it is fitting that the recording or recapturing of such lives is termed biography—life writing—for these were lives that were known.

In my contribution, I explore the afterlives of animals that begin when they become the focus of human interest and attention in their natural habitat and when a hunter decides that a particular animal is huntable and killable. What I am interested in here is what the hunter seeks in the process, in the relationship with a wild animal, that brings about the commencement of its afterlife and gives value to it as a trophy. Although each animal, now represented as a trophy, might have been alive for many years before its fatal encounter with a hunter, the biography of that wild life is not captured in the trophy, for it is unknown. They had individual lives, but they had no specific, individual biographies prior to the moment of being selected for death by the hunter. With their deaths, each animal becomes a significant individual, each becomes important for what they once were, and how they once were, as they were hunted. At the end of their life, a story can be told about that life; they can have a biography, albeit a short one, and this biography as an afterlife becomes inextricably tied to the autobiography of the hunter.

The afterlives of most of the animals discussed in this volume have two aspects—how they lived when alive and how they lived on after death. The afterlives of hunted animals that have become trophies are somewhat different. How they actually lived, prior to being hunted, is irrecoverable—although it is perhaps imaginable. What is significant is how, at the end of their lives, they were brought into a relationship with a particular hunter. This relationship was not willed or sought out by the animal; but neither were any of the other relationships described by others here. It is, however, the creation of a particular relationship, through the gradual engagement of the hunter with the hunted, that is celebrated in the afterlives of hunted animals. This is not a celebration of how the animal lived; rather it can be interpreted as a celebration of the process of how the hunter was able to bring about its death. Such a relationship is a deeply personal one between the hunter and the hunted, and it is appropriate, or perhaps culturally significant, that the taxidermied trophy does not begin its cultural life in a public space but rather in the home, the private space, of the hunter. The trophy also begins a life in a collection, but in a collection that is very different from a collection of taxidermied animals in a museum.

What unites the animals in a hunter's collection is not that of any scientific or taxonomic ordering; rather it is that the collection is linked to the autobiography of their hunter. Fundamental to the creation of such a collection is that all the individuals have been hunted by that hunter, and they have become his or her animals. What is crucial here is not that trophies are possessed but how they have become possessed by the hunter; the manner of their acquisition is inextricably tied to their hunter. As I explore in this essay, it is not so much that they were *killed* by the hunter that is significant but that they were *hunted* by that hunter.

The argument I seek to develop here is that hunting trophies as material objects are primarily markers of what developed between the hunter and the hunted, the experiences of the hunter prior to the death of the animal. Their significance does not principally reside in what they are in the present (although they do have such significance in terms of their taxidermied qualities) but rather in their power to evoke the past, a past that can be relived by the hunter and by only that hunter. Such a collection is more than an assembly of individual trophies; it is also a complex site of recollection at which animals that had no connection with each other when alive are brought together because of the manner of their deaths.

Part of my argument is concerned with notions of authenticity as they relate to how hunters regard their trophies and the trophies of others.[1] There are two key elements to notions of authenticity here. The first relates to what occurred between the hunter and the hunted in the habitat of the hunted. This notion of authenticity, that the authenticity of the trophy should relate to the authenticity of the hunting process and, at its close, how the death of the animal was brought about, is the most complex of the two and is given greater attention here. This is an authenticity of a particular trophy that can be ascribed to it only by the hunter. Such authenticity does not reside in the object itself, and it cannot be authenticated by an outsider; it can be ascribed to it only because what is authentic about it exists in the past and in memory. The second aspect of authenticity relates to the physical object that is the trophy and how that object, through the taxidermist's craft and art, captures, re-creates, and fixes the lifefull quality of the animal prior to its death.[2] This quality of authenticity is something that can be, and is, judged by aficionados of the art of taxidermy, for it resides in the physical presentation of the trophy as a representation. My overarching claim is that to understand fully what hunting trophies represent, and why hunters want them, it is necessary to pay close attention to that "represent" from within, from the perspectives of hunters, rather than simply to impose meanings on them from outside.

In his exploration of trophies in the world of hunting, Ted Kerasote refers to one hunter as "heading to the other side of the world, again to the hunting fields, making memories with his rifle," and to another who, if successful, "will enshrine the trophy in a place of honor."[3] In a recent conversation, a Spanish hunting friend explained to me that each of his mounted heads was *un recuerdo*—a term that refers to reminiscence, to memory, and to a material souvenir. The notions of the enshrinement of a venerated object and to a souvenir embodying a special memory suggest to me that the taxidermied, or otherwise prepared, parts of hunted animals that are brought into the homes of their hunters are more complex objects, and partake in more complex processes, than are recognized in the simple critiques of such objects as material celebrations of a hunter's power, domination, and triumph.

## AUTHENTIC HUNTING — AUTHENTIC TROPHIES

The Spanish philosopher José Ortega y Gasset described hunting as a "subtle ritual," and in this essay I explore how the personal hunting trophy is a subtle object that emerges from that ritual process.[4] My interest here is in how the material, in the form of the taxidermied body, relates to memory. This connection is intimately related to the notion of a vestige, that is, a "mark, trace, or visible sign of something no longer present or in existence."[5] The taxidermied animal in a hunter's collection is, at the material level, a vestige of the animal, but at a more complex, experiential level, it is also the vestige of a relationship formed during the process of hunting; one that now resides in the home and the memory of the hunter. However, such relationships cannot be immediately discerned in what remains of the body of the animal itself. Without the enlivening presence of the hunter, such objects are mute and can probably be viewed only as dead animals. The significance of the trophy, the fact that it is displayed at all, is revealed only when hunters speak about them. Not all are valued equally or in the same manner. Some may be trophies in the original, military, sense of spoils of battle and symbols of both victory and defeat on the hunting field. Some may be valued in and of themselves as the biggest, rarest, or the most difficult to hunt. Others, the focus of this piece, are sites of memory that both invoke and evoke the hunter's journey out to the animal and the return with it. Not all hunters make the journey in the same manner, for the same purpose, or for the same experiences. My focus here is on one type of hunter, one set of orientations to hunting, and the trophies that are created as a result.

In an early study of types of hunters, Stephen Kellert made the important point that modern leisure or recreational hunters do not form a homogeneous group, and in this study he distinguished three categories of hunters: utilitarian/meat hunters, dominionistic/sport hunters, and nature hunters. He suggested that "utilitarian/meat hunters appeared to perceive animals largely from the perspective of their practical usefulness. [They] viewed hunting as a harvesting activity and wild animals as a harvestable crop not unlike other renewable resources." For the dominionistic/sport hunters, "the hunted animal was valued largely for the opportunities it provided to engage in a sporting activity involving mastery, competition, shooting skill and expressions of prowess." For the nature hunters, "the desire for an active, participatory role in nature was perhaps the most significant aspect of the nature hunter's approach to hunting. The goal was the intense involvement with wild animals in their natural habitats. Participation as a predator was valued for the opportunities it provided to regard oneself as an integral part of nature."[6]

Kellert alerts us not to read these as exclusive categories. Although hunters tended "to be orientated toward one *primary* attitudinal relation to hunting," they were "typically characterized by more than one attitude."[7] I suggest that what is important for understanding hunting is that it is not simply a matter of "attitudinal relation" but also a matter of attitude informing and guiding hunting practice on different occasions and for different purposes. Rather than using these types as a classification or categorization of *hunters,* I find them more useful for thinking about *modes* of hunting, or as different *orientations* to hunting in any particular hunting event. Different modes or orientations may come to the fore on different occasions. So, in the context of hunting trophies, a trophy hunter might also espouse the values of a nature hunter, and a nature hunter might seek a trophy on a particular occasion, while their primary orientation is to achieve an intense involvement with the natural world through hunting. My focus here is on hunters whom Kellert would classify as nature hunters and on the trophies of their hunts that they display in their homes (see fig. 1).

Underpinning the practices of such nature hunters is a fundamental belief in, and adherence to, what has been termed the ethics of "fair chase." This ethical approach to hunting is constructed from a set of beliefs about the appropriate and essential relationships that should obtain between the hunter and the potential or actual prey. Key elements of this relationship are that the hunted animal should be a naturally wild animal, that is, not an animal specially bred and raised for hunting; that it is not restricted in any way such that it cannot escape the hunter; that it should not be pursued and killed from a vehicle; that hunters should voluntarily restrict their ability (particularly in terms of the use of technology) to hunt an animal; and that, at the moment of taking a shot, the hunter must be confident that the shot will be a lethal one. Ethical hunting is summarized in Jim Posewitz's declaration: "Fundamental to ethical hunting is the idea of fair chase. This concept addresses the balance between the hunter and the hunted. It is a balance that allows hunters to occa-

FIG. 1. "A Private View," 2010. (Photograph by Antonio Marcelo Herrera)

sionally succeed while animals generally avoid being taken."[8] Its "occasional" occurrence is crucial. For all the hunters with whom I have worked, the idea that a kill can be guaranteed during a hunting event immediately suggests that there is something artificial about it, that somehow the odds will be stacked in favor of the hunter, and that the hunting is unlikely to be what, for them, is true hunting.

For nature hunters, fair chase is authentic hunting, and for them, the hunted animal should be converted into a trophy only if it has been hunted in an authentic manner; without that process, the trophy has no value. As Arjun Appadurai comments, citing Georg Simmel, in his exploration of the nature of commodities in cultural perspective: "Value ... is never an inherent property of objects, but it is a judgement made about them by subjects," and "the difficulty of acquisition, the sacrifice offered in exchange, is the unique constitutive element of value."[9] The "difficulty of acquisition" is important here, as is that "judgement made about them by subjects." If the opportunity to kill an animal has been made artificially easy, then there is no challenge for the hunter, and, for the hunters I know, this is simply not hunting. For such hunters, the *nature* of the process is essential, and the trophy produced as a result comes to represent the experience of that particular hunt. Two examples might help to illustrate how the notion of an authentic trophy should relate to authentic hunting.

As part of the research for his study of the world of trophy hunters, Ted Kerasote reports on how one of the hunters explains the significance of a mounted head of an aoudad, or Barbary sheep:

> "If you go into anyone else's trophy room," he says, "and see this animal, you will *never EVER* see a specimen so small. Its horns are nineteen and a half inches long. I'm sure some of the people you've talked to have aoudads of twenty-nine to thirty inches. This one was shot in the Nubian desert. Theirs were shot in Texas or New Mexico, on a ranch, behind a fence. Anyone who comes in here, and knows about sheep, says, 'Holy Christ! Look at that.' Because they know where it came from. I will *never* go shoot one of those thirty-five inchers in Texas because I know what it took to do this."[10]

Another of his hunters criticizes a fellow hunter who, on a trip, was so obsessed with the size of a trophy, rather than the experience that came before the kill, that he was disappointed with the outcome as soon as he measured the horns of the dead animal—"That poor son-of-a-bitch is lost. He's lost the essence right there."[11]

The second example comes from my Spanish hunting friend who regards his trophies as souvenirs. We were sitting in his trophy room, and he was trying to explain to me that it was essential for him to hunt with "dignity and respect." As an example of the very opposite of that, he said, "and that is why the elk is on the stairs and not in here." We went to the stairs that led to the room, and he pointed out the skull and the antlers of an elk. He explained that although these were antlers of an international medal size, it was the most shameful trophy he had and that is why it was impossible for him to have it in the trophy room. The reason that it was shameful was that he shot the animal on a huge, enclosed ranch where the animal had no chance of escape. He had shot it, he had killed it, but he had not hunted it, and so it was not an authentic trophy. He admired the huge antlers as "a work of natural art," but it reminded him of what he did as a young hunter and what he would never do now.

Ethics and authenticity, how one believes one ought to hunt and how that has informed how one actually hunts at the moment of hunting, are the personal responsibility of the individual hunter. The trophy is something that the hunter personally awards himself or herself in recognition of that process. In this sense, a hunting trophy is very different from a trophy acquired in a sporting event. Such a trophy is awarded by others as a result of an achievement, in terms of recognized standards and rules, that is witnessed by others. The process of hunting rarely has witnesses, other than perhaps by guides, and the only judgment of achievement is made by the hunter.

## SOUVENIRS — OBJECT AND MEMORIES

In her essay "Objects of Desire," Susan Stewart suggests that the "capacity of objects to serve as traces of authentic experience is . . . exemplified by the souvenir."[12] It is in the manner of acquisition and in its relationship to the hunter, in terms of authentic experiences, that the hunting trophy might be interpreted as a souvenir—a material object from elsewhere, and from another time, that is imbued with meaning and memory when brought home. Objects can become souvenirs in a variety of ways. Apparently insignificant objects, such as a ticket stub from a favorite concert, a wine bottle from a memorable occasion, a child's first clothes, might be kept as souvenirs, as "tokens of heightened moments."[13] Souvenirs also come in the form of specifically manufactured souvenir objects, sold at key tourist sites, that represent those sites through a reproduction of signs and images—models of the Eiffel Tower in Paris, copies of the Statue of Liberty in New York—or, of a slightly different order, rugs from Istanbul, masks from Venice, bottles of maple syrup from Canada, and spices from an Indian market. For a souvenir to be authentic, it must be obtained in the place that it represents by the visitor who has been there. While in the place they represent, souvenir objects are not yet, fully, souvenirs. The rows of masks and model gondolas on shelves in a shop in Venice are not souvenirs for the shopkeeper; they are merely goods for sale that must await new owners, from outside Venice, to become souvenirs. In the place they represent, objects only have the potential for evocation of the experience of that place, for the purchasers are still within that experience. The work of the souvenir begins only when it is removed from that place and when experience is replaced by memory. When it is brought home, the unindividuated object of the souvenir shop becomes a personal memento of the person who obtained it. An object is not a souvenir because of some essential quality it has, or because of what it is; it only *becomes* a souvenir when it enters a relationship with a person. But such objects do not speak for themselves; they must be activated, brought to life, either through the thoughts and reminiscences of the owner or when the owner explains their significance to others. As Stewart argues: "We do not need or desire souvenirs of events that are repeatable. Rather we need and desire souvenirs of events that are reportable, events whose materiality has escaped us, events that thereby exist only through the invention of narrative."[14] As I later suggest, hunting trophies come alive through the reveries or narratives of the hunter—they become the sites of memory and the focus for stories and reminiscences of hunting.

Hunting trophies have "souvenir qualities" in that they are tokens of heightened moments, sometimes emblematic of the place where the animal was hunted, and are objects that are brought back from elsewhere to be displayed as mementoes of experience in that other place. Where they differ from other souvenirs, however, is that authentic trophies cannot be purchased; they must be created by the hunter. There is no such thing as a generic hunting trophy as there are generic souvenirs, and no hunter would display as a hunting trophy an animal that she/he had not personally hunted and killed; a trophy shot by someone else would have no significance for another hunter. No hunter I have spoken with would have a trophy in his or her home that she/he had not hunted and killed. So, whereas souvenir objects can be purchased by one person and given to another as part of the ritual processes of tourism, travel, and vacationing, hunting trophies do not enter such gift exchanges. The act of their creation as souvenir objects begins in the process of hunting itself and continues through the stages of converting a dead animal into a mount; a process that converts an impersonal, living animal into a representation of, or a re-creation of, one that has a unique and personal relationship with the hunter.

As Stewart notes, "The souvenir is by definition always incomplete." The wine bottle no longer contains wine, the child's clothes are no longer worn by the child who made them significant, and the model of the Eiffel Tower bears no relationship to the size and grandeur of the object to which it refers. Stewart also refers to the incompleteness of the souvenir object as "metonymic to the scene of its original appropriation"; it has a part/whole relationship with it.[15] In this sense, the hunting trophy is metonymic in two ways. At one level, its reference is to the particular hunt out of which it was produced, but it is also metonymic of the specific animal to which it refers. Hunters do not seek to convert the entirety of the hunted animal—flesh, organs, bones—into a trophy.[16] In order to represent any animal, most of it must be discarded, and only those parts—skin, skull, hooves, claws and teeth—that can be preserved from biological deterioration are kept to replicate the whole. Although the biological must be rendered inert, taxidermy is not concerned with the preservation of natural objects, dead bodies. Taxidermic objects are not dead animals preserved; rather they are cultural objects created through craft. However, it is the dead animal and the entire body that is first subject to conversion into a souvenir.

The lethal shot and the engagement of the hunter with the now dead animal mark the beginning of two kinds of reliving that take that particular hunt into the past and into the future. The hunter begins to relive the hunt as he speaks about it with colleagues and guides who might be present. In my experience, once a hunt has moved from an unfocused scanning of, and an unfocused movement across, a landscape to a focused movement toward a huntable animal, the hunter is so enveloped and absorbed in the immediacy of the process that words and reflection are impossible. With the death of the animal, the hunter begins to re-create the experience for him or herself and for others. The hunter also begins the process of converting the hunted animal into a souvenir, and that first quality of a souvenir is the photograph.

The photograph of the hunter posed with the dead animal shares many elements with a tourist's souvenir photographs created at a place of significance in order to record being there for future display and reflection. A significant feature of touristic photographs, in which the tourist is present, is the relationship between the person and the site. In the act of sightseeing, tourists look outward, toward the sight of interest. Tourists do not usually asked to be photographed with their backs toward the camera in the act of looking at a work of art in a gallery or when walking around a church. The act of recording a person at a place of significance involves a pause and a reversal of the angle of sight. In such photographs, the site becomes the background; the tourist faces away from it, toward the camera and outward to the future. Something similar occurs in hunting photographs. Most hunting photographs do not record the hunter in the act of hunting; they record the end of hunting. While hunting, the hunter is absorbed in looking toward the animal (a parallel with the tourist who looks outward to a site/sight of interest), and it is not part of the process to be concerned with being seen. In capturing the end of hunting, the angle of view is reversed and the hunter looks toward the camera and not toward the animal. As with tourist photographs, trophy photographs involve a pause and a concern with pose.

In all hunting events I have witnessed, the posing of the hunter with the dead animal is very carefully managed because, as the hunters have commented, this is a moment that is unique and unrepeatable; it captures being there. In my experience, the posing of the animal is not, as others have argued, to pose the animal in the most lifelike position; hunters are not ashamed of their actions and do not seek to simulate life out of the death they have brought about.[17] Donna Haraway's comment on a hunting photograph in which the hunters "look directly at the camera in unshuttered acceptance of their act" would certainly apply to all the hunting pictures I have watched being created.[18] Hunters are clearly recording their connection with a dead animal, not a simulated live one. They pose with it, and they often include the weapon by which they brought about that death.

Rather than seeking a lifelike pose, the body is posed to reveal the physical qualities that the hunter most admires in the particular animal, for example, its size, the quality of its horns or antlers, the size of teeth or paws, or its general beauty. What is also of great significance is creating the connection between the hunter and the hunted animal for this animal is now a particular and a personal animal; it is the animal of the hunter. Such a photograph is very different from a tourist safari photograph depicting a wild animal; apart from the obvious fact that the tourist photograph is usually of a living animal. Here images of a kudu, a lion, or a water buffalo are of specific animals, but they are also of an animal in general, an animal with which the tourist has only a fleeting encounter through the camera. A tourist might display at home a photograph of a wild animal, but there is little that can be said about that specific animal because there has been little engagement with it. The hunting photograph is a record of a different order for it is the record of a relationship that will be recounted in great detail later. The narrative of a hunting photograph involves accounts of how it was to be there, the difficulty or ease of approaching the animal, how and why that particular animal was selected, and how it was to take the shot.

The photograph marks the end of the fleshy, hunted animal and the beginning of the process of re-creating and reenlivening it in a cultural form. In my experience, and from my reading of the literature of hunters, nature hunters have a highly developed aesthetic appreciation of the animals they choose to hunt, and their accounts of hunting involve descriptions of the pleasures of seeing the beauty of these animals and their ways of being in their natural environments. A series of aesthetic appreciations, considerations, and judgments also draws the hunters' attention to an individual animal on which to focus the sights of their weapons in order to make that animal theirs. After the death of the animal, a new aesthetic enters; a personal aesthetic that guides the process of recapturing and fixing a sense of what that animal once was and how that animal was seen and experienced by the hunter.

At the point of preparing the hunting trophy, most hunters turn to the services of a professional taxidermist because, although they have a vision of what the mounted trophy should look like, they do not have the craft and artistic skills to realize this vision. Here I return to the comment that the trophy is a metonym of the hunted animal for it can never be a complete animal. In an important and significant sense, the taxidermied mount is only a superficial animal; it is only, literally, skin-deep, but that surface must be crafted to convey a sense of the whole. In this sense, each can also be considered as simulacrum, as "a thing having the appearance but not the substance or proper qualities of something," that is attempting to convey those proper qualities of the original.[19] Hunters and taxidermists with whom I have spoken have commented that the ideal hunting trophy should be of the whole skin to re-create a whole body; otherwise the trophy will always be partial and impoverished. However, most hunters do not seem to have the space to display all their trophies as complete bodies, and, in the collections I have seen, heads are more common than complete bodies. As with representations of people in portrait painting, photographs, and sculpture, it would seem that it is the head and face of the hunted animal that best conveys the essence of that animal, or can memorialize that animal, the part that is the most expressive and individual.

All hunters have told me that expressiveness is what they seek in a mounted trophy. A fine trophy head should communicate the spirit and essence of the animal. The pose in which it is mounted should also be natural, lifelike, and communicate lifefullness. As one taxidermist put it to me when discussing how he was going to mount a full leopard skin, "If it is not done well it appears that the leopard is not there." Through his skill he had to bring about the animal, to give it presence, and to make it present in as close a way as possible to the animal it once was. He also said that he refused to take on commissions when the client wanted something prepared in what he, the taxidermist, regarded as exaggerated, unnatural, overly aggressive, or demeaning poses. I have frequently heard criticisms of trophy mounts in these terms or in terms of being lifeless or as representing stereotypical cultural images of animals rather than capturing the essence of how the animal normally is in the wild. I questioned one friend about why he had chosen to have his brown bear mounted on all fours rather than having it rearing on two legs to show its full height and the size of its paws. He responded, "Because that is how they usually are—that is how it was when I saw it; bears do not go around walking on two legs, snarling aggressively." Another friend, commenting on how poorly the natural wolfness of wolves was captured in two taxidermied snarling heads, said, "This male is too exaggerated, too big, too wild, and with this female they have given her the teeth of a vampire!"

At the beginning of this essay, I quoted Kerasote writing about a hunter enshrining a trophy. From my reading and research, I do not think that it would be appropriate to think of an enshrinement in the sense of honoring the dead animal as such, although it perhaps has elements of that. It also has an element of monumentality in the sense of a monument erected to honor and remember an individual or an event, although here, rather than the normal collective and public monument, it is a personal and private one.

Hunters tell me that when they sit and contemplate their collection, each trophy unlocks their memory, and they return, in their imagination, to how it was to be there on the hunting journey. They tell me that they remember the pleasures and discomforts of travel, of being in different places, of the scents, sights, and sounds around them. They remember how their bodies responded to the weather and the terrain and how it was to inch themselves slowly and carefully toward their chosen animal. This is also how they narrate the significance of a trophy to those whom they invite to see their collection. "That day it was bitterly cold. We left at dawn, and we were soon wet through; it took us five hours of hard climbing to find the herd and then we had to crawl." The description of being there and the approach is usually rich in detail, but the moment of killing, unless it was a particularly skillful shot, is hardly elaborated. The kill is what ends the hunt; to have hunted is the thing, to have been able to get into a position to take the shot. Hunters also have narratives of the trophies they did not obtain because the animal evaded them. These, too, are narratives of how it was to have hunted, although, in the case of failed trophies, these are sometimes tinged with regret and frustration; but that too, in their terms, is also hunting and to have hunted.

The hunting trophy, without the hunter, is nothing more than a lifeless artifact. With its hunter, it has a memory in which it can live. I believe that this artifact can best be interpreted as an enshrinement of a process, an engagement and a relationship that connects the hunter, the terrain, and the hunted animal. If the hunter has, in his or her terms, hunted properly and well, then the hunting trophy memorializes that hunting. As the hunter and philosopher Allen Jones comments on his style of nature hunting: "Hunting is not an attempt to take possession of the animal, as so many hunters have argued. If that were the case, the attempt is doomed to absolute failure. I cannot possess the elk. It's dead. But I can possess the memory of it in the moment before I killed it, which is enough. It's more than enough. What I have brought away is my relationship to the animal, not the animal itself."[20]

## NEW AFTERLIVES

But this personal relationship cannot last forever. One side of the relationship ended when the animal was killed, but the hunter, too, must die, and with his or her death, the immediacy of the particular and intimate relationship between the animal and the person comes to an end. The first afterlife of the animal represented in the hunting trophy is a temporary condition that cannot continue longer than the life of the hunter. But new afterlives are possible.

One possibility is that the family of the hunter might preserve the collection intact, and it might continue to live in the family home, perhaps for generations. Here I would speculate that the animal in the trophy becomes inert, a dead animal, for what it was when alive and the experiences that brought about its taxidermied condition cannot be remembered. What can be remembered, perhaps celebrated through the trophies, is the life of the hunter, although that, too, must fade as the hunter slips from being a person known to an ancestor known about. Through time the collection might, perhaps, be transformed from being a shrine to animals and hunting experiences to a shrine to the hunter or to a collection of heirlooms.

There is another possible afterlife of hunting trophies, one that takes them out of the private and into the public places and public processes explored by colleagues in this volume. Hunting trophies might be given to, or otherwise acquired by, natural history museums. Once again, though, the animal that was is disconnected from the hunter. In the museum, the trophy ceases to be a personal trophy, something of significance in terms of the processes through which an animal became preserved in this form. Rather it becomes an impersonal representative of a species.

Placed among hundreds of similar bodies, it might simply be ignored, but perhaps as a public object, the taxidermied mount enters a more complex, multifaceted afterlife in a museum. It might be seen, noticed in passing, by thousands of visitors over the years; it might be more closely observed and contemplated by some who find themselves drawn to it; scientists might study it; curators might use it to tell a range of stories, and others might analyze it in terms of the craft and art of taxidermy from a particular time and a particular place. Some, as is the case with the histories of animals discussed in this volume, may attract the attention of researchers who, through an exploration of their particular and individual history, draw them out of the obscurity of being mere species representatives and reconnect them once more with those who hunted them and who, originally, marked them out as worthy of having afterlives.[21]

## NOTES

An earlier, and more general, version of ideas expressed here appeared as "Living with Dead Animals?: Trophies as Souvenirs of the Hunt," in *Hunting — Philosophy for Everyone: In Search of the Wild Life,* ed. Nathan Kowalsky (Malden, Mass.: Blackwell, 2010), 107–17.

1. Here I do not have the space to explore the complex philosophical issues relating to authenticity and objects, but for a useful overview of issues relating to authenticity and experience, I would refer readers to Ning Wang, "Rethinking Authenticity in Tourism Experience," *Annals of Tourism Research* 26 (1999): 349–70.

2. I use the term "lifefull" rather than "lifelike" in this essay to capture the sense of the quality that hunters seek in a trophy. Lifefullness is the quality that should emanate from within the object and communicate outward. It should be individual and specific to that animal. "Lifelike" suggests a lesser quality of mere appearance—an approximation to, or an imitation of, life. "Lifefull" also works nicely against its opposite, "lifeless."

3. Ted Kerasote, *Bloodties: Nature, Culture, and the Hunt* (New York: Kodansha International, 1993), 87, 85.

4. José Ortega y Gasset, *La caza y los toros* (Madrid: Revista de Oriente, 1968), 119.

5. *Oxford English Dictionary.*

6. Stephen Kellert, "Attitudes and Characteristics of Hunters and Antihunters," *Transactions of the North American Wildlife Resources* 43 (1978): 414, 417, 415.

7. Ibid., 413.

8. Jim Posewitz, *Beyond Fair Chase: The Ethic and Tradition of Hunting* (Helena, Mont.: Falcon, 1994), 57. See also Garry Marvin, "Challenging Animals: Purpose and Process in Hunting," in *Nature and Culture,* ed. Sarah Pilgrim and Jules Pretty (London: Earthscan Books, 2010), for a fuller account of the nature of hunting from the perspectives of nature hunters.

9. Arjun Appadurai, ed., *The Social Life of Things: Commodities in Cultural Perspective* (Cambridge: Cambridge University Press, 1988), 3, 4.

10. Kerasote, *Bloodties,* 97.

11. Ibid., 126.

12. Susan Stewart, *On Longing: Narratives of the Miniature, the Gigantic, the Souvenir, the Collection* (Durham, N.C.: Duke University Press, 1993), 135.

13. Dean MacCannell, *The Tourist: A New Theory of the Leisure Class* (New York: Schocken, 1989).

14. Stewart, *On Longing,* 135.

15. Ibid., 136.

16. It might be possible to make the argument that the hunter who takes home the meat of a hunted animal and who then butchers and preserves it for later eating is also creating a souvenir object and that the eating of such meat might be a souvenistic event.

17. Linda Kalof and Amy Fitzgerald, "Reading the Hunting Trophy: Exploring the Display of Dead Animals in Hunting Magazines," *Visual Studies* 18 (2003): 112–22.

18. Donna Haraway, *Primate Visions: Gender, Race, and Nature in the World of Modern Science* (New York: Routledge, 1989), 34.

19. *Oxford English Dictionary.*

20. Allen Jones, *A Quiet Place of Violence: Hunting and Ethics in the Missouri River Breaks* (Bozeman, Mont.: Bangtail Press, 1997), 104.

21. For a fascinating account of how two artists sought to discover the provenance, and recuperate the histories, of taxidermied polar bears in the United Kingdom and to then reexhibit the bears in a new venue, see Bryndís Snæbjörnsdóttir and Mark Wilson, *Nanoq: Flatout and Bluesome: A Cultural Life of Polar Bears* (London: Black Dog, 2006).

GEOFFREY N. SWINNEY

# An Afterword on Afterlife

Wildlife occurs within ecosystems, while the after-lives accounted for in this book are enacted in and through (human) social systems. In the museum, it is the visitor who breathes new life into objects, and, in the case of representations of once-living organisms, that "new life" is what we have classed as its afterlife. The preceding essays recount particular kinds of narratives and thereby produce and define particular kinds of afterlives—principally sequels to some measure of celebrity status acquired by an individual animal, or imposed upon it, while it was alive. These are the animals that gain entry to celebrity listings such as the *Animals' Who's Who.*[1] Further, the essays focus mainly on afterlives conducted through elaborate material reconstructions exhibited in the front-of-house spaces of museums. In this afterword, I consider further the broad themes, transitions, and meanings which pervade and structure this volume, but I seek also to extend the parameters beyond those addressed in the preceding essays. I suggest that the museum menagerie is animated and vitalized through a broader range of afterlives, that these are not restricted to vertebrates, not necessarily associated with celebrity, and extend through a variety of spaces of the museum, and beyond. To expand the concept of afterlives of museum zoological specimens, I develop two interrelated strands, both substantially concerned with matters of practice. The first addresses the ways in which animals are (epistemically) reconstructed in the museum menagerie. It focuses on the creation of identity and individuality, including celebrity identities, and on the spaces of the museum which the menagerie occupies. The second strand of discussion considers further the transition from carcass to specimen, and centers on the technologies of preservation and material reconstruction.

## THE MENAGERIE ON DISPLAY

The afterlives explored in the preceding essays have generally been those enacted by a remnant of the animal's body, preserved and fashioned into a model, which through its pose connotes that particular individual animal in life: either its external appearance or its (re)articulated internal skeletal structure. Such models are themselves reliquaries, each fashioned into the appearance of some aspect of the celebrated individual that it memorializes. Of course, not all celebrity afterlives are a sequel to a celebrated life; a few animals acquired celebrity only in afterlife. Hannah Paddon notes the example of the giraffe which was shot and mounted as a hunting trophy—a fine representative of its species—and which gained an individualized identity, its mascot status, and the name Gerald, only once placed on display in the Royal Albert Memorial Museum, Exeter, United Kingdom.

Exceptions such as Gerald notwithstanding, while they were alive, most of the animals considered in this volume enjoyed (or endured) a close, and often protracted, association with people. Most did not share domestic space, but rather passed at least part of their lives as a component of a zoo, a circus, or a menagerie. In captivity, each of the animals became an adjunct to human society rather than being a member of a herd, flock, or pod. Anthropomorphized, each was adopted as more than merely a representative of its species—expressed in language by the use of "he" or "she," "who" or "whom," rather than "it" and "which," and most especially by the bestowal of individual identity, often through the allocation of a human-type name. As Richard Sabin recounts in "The Thames Whale," even the short-lived appearance of a northern bottlenosed whale (*Hyperoodon ampullatus*) in the River Thames in the center of London prompted the British press to bestow upon it a variety of names including Prince of Whales, Pete the Pilot, Whaley, Willy, and Wally, and to situate it within popular culture by reference to "Celebrity Big Blubber."[2] Similarly, most of the other animals discussed in this volume were adopted by the media and the popular culture of their day, thereby gaining celebrated status—itself a form of "cultural memory that speaks beyond individual experience."[3] Preservation and (re)presentation of some conspicuous fragments of the body, modeled to mimic the external appearance or the internal structure of the living animal, provided a means for the animal's celebrity to transcend death, thereby affording new forms of engagement in human society.

Even in their premortem existence, these animals were appropriated and reconstructed in *our* image. They were anthropomorphized and fashioned to embody human emotions and values; constructed as creatures with which we were able to empathize. Death allows such roles to be consolidated, and the postmortem reconstruction of an animal is both material and epistemological. Preservation and reconstruction divest the animal of those aspects of its animality—its beastliness—which serve to remind we humans of our own biology and of the beast within. Reconstruction, involving processes of flensing and cleansing, absolve the animal of the necessity for such base functions as urination, defecation, and the overt signaling of sexual receptivity and eagerness to mate.[4] It removes from it all ability to spit, bite, slobber and drool, kick, scratch, or generally run amok in behaviors which communicate nothing in human repertoires of understanding other than chaos or cacophony. Further, the preservation process deprives it of the ability to rot and create stench, other life-cycle processes which we humans efface in relation to our corporeal selves. The animal as mount is the cleansed and purified "ennobled" beast, immaculate in behavior and manners. It is an animal which is chaste and without vulgarity, which transcends the ephemerality of biological life and is incorruptible by age, sickness, death, or decay. Even skeletons stand effacing death, cleansed of flesh (and its frailties), bleached, and mounted on their own metal "skeleton" or armature of rods, wires, and bolts. These are not the prone and scattered bones of decay, but the structural and sculptural inner core of the animal, a revelation of the mysterious heavy-engineering of the once-living creature. The materially reconstructed animal is immune to pain, free from our ability to inflict upon it (further) mental and physical cruelty or violence. Any guilt or concern about the animal's welfare—its confinement, its isolation from its own kind's social structures, and the enforced company of humans—troubles us no longer; the animal is "at rest," while simultaneously prepared to be continually at our behest. Modeled in glass, its eyes are incapable of any disturbingly accusatory stare.

The preceding essays might seem to suggest that there are certain characteristics shared by the animals which gain celebrity and the associated rite of passage into afterlife. All are vertebrates; most are mammals. This bias reflects and contributes to a tendency, evident in a variety of popular literature and also in some technical publications, for "animal" to be used as if it were synonymous with "mammal," thereby relegating the rest of the animal kingdom to subaltern status.[5] Another shared characteristic is that all belong to species which attain a moderate to large body mass—a disproportionate number are gigantic—suggesting a correlation between size and charisma. As Samuel Alberti discusses in his essay on Maharajah the elephant, large body size, so frightening if the animal is running amok, has charm when the creature is modeled and reconstructed as a gentle giant. Only exceptionally do small animals gain celebrity. The ability to speak helps: in life, Sparkie Williams (1954–1962)—a budgerigar (*Melopsittacus undulatus*) who reputedly had a vocabulary of more than five hundred words and phrases—advertised bird seed and released recordings on major record labels (recently reissued on CD); in death, he inspired musical experimentation and earned a place on display in the Hancock Museum, Newcastle upon Tyne.[6] Equally exceptional is the granting of celebrity to lower vertebrates, although a few cold-blooded giants have become celebrated or notorious. These include some large fishes, such as the female carp Benson, targeted by anglers, the forlorn giant tortoise Lonesome George, and the feared crocodiles Gustave and Sweetheart—the latter now in afterlife in the Museum and Art Gallery of the Northern Territory, Darwin, Australia.[7] Very few invertebrates gain celebrity and individual human-type names, although a sea anemone (*Actinia equina*) known as Granny was deemed worthy of an obituary in a national newspaper in the nineteenth century, and the death of Paul, an octopus (*Octopus vulgaris*) who reputedly predicted the results in the 2010 soccer World Cup, recently prompted a flurry of media notices.[8]

The higher vertebrates (mammals and birds) typically have insulating body coverings of nonliving keratinaceous material in the form of fur or feathers. These structures, when dried, are resistant to decay and make the hides of these animals readily preservable in a manner which superficially resembles their appearance in life. Similarly the bony skeletons of these animals are easily preserved by desiccation.

Celebrity depends upon the individual animal maintaining a recognized and recognizable presence in the museum menagerie. These identities are maintained, not by the representation of the animal alone, but through practices of material representation working in concert with practices of documentation. Each animal travels into and through afterlife accompanied by identity papers. It is the combination of technologies for preserving the material remains of animals together with those that capture and preserve knowledges about the how, when, and where of those animals' trajectories to (and within) the museum that enable afterlives. Without such information, Dolly—the first mammal to be cloned from an adult cell and the world's most famous sheep—would be indistinguishable from the rest of the flock; Sir Roger, disguised by false tusks, might go unrecognizable even to those who knew him well in life; and Maharajah's skeleton could pass for that of any other adult male *Elephas maximus*.[9] But such celebrities are not alone in having individual identities imposed on them. Indeed, the creation or maintenance of individual identities is central to the operational practices of a museum. Typically, these involve assigning each object a unique identifier, usually a numeric or alphanumeric "accession number" or "registration number." This identifier associates the object to documents relating to its pre-museum existence and facilitates subsequent documentary practices—labels, accession books, registers, location indexes, loan records, conservation reports, display labels, research publications, exhibition catalogues, museum guidebooks, and a variety of other paperwork—which track its museum career.[10] As Stephen Jay Gould suggests, within natural history collections these rites of ascribing individual identity, along with the processes of preservation and representation, are fundamental to the construction of afterlives: "We have devised peculiar rites for these creatures in natural history museums—inscriptions in ink on bone, chemical baths to render them translucent, systematic placement on shelves, often in the dark, often in shrouds of dust or moth crystals. I think of these treatments as forms of burial, but I think of the animals as expressing, in various ways, life after death."[11] It is through their individual identities that the animals in the museum menagerie do their work of creating understandings of the natural world. Yet, I argue, this work is by no means restricted to celebrities or to vertebrates; instead it is open to all inhabitants of the menagerie, regardless of their taxonomic place within the animal kingdom. It includes a multiplicity of voucher roles, which may include acting as proxies for environmental local or global conditions in times past; illustrating the extent of intraspecific diversity; attesting to the physical and epistemological journeys of explorers and scientists of all kinds; and/or providing the material objective evidence for past subjective taxonomic judgments. One particular category of voucher, the type specimen, serves the function of being the objective reference within the international system of zoological naming—the process of appellation through which a species becomes "known to science."[12] For John Huber, the role of voucher specimens of all kinds, providing as they do the ability to revisit the material basis of past subjective judgment, is fundamental to the process of science and is crucial in making natural science "scientific."[13] In addition, as the contributions to this volume demonstrate, interrogation of the documentation associated with individual specimens may also be revealing of their passage to and within the museum—their career trajectories in life and in afterlife.[14]

Gould's description of dust-shrouded dark spaces, and his use of "burial" as a metaphor for the process of museum storage, suggests that he was concerned with spaces other than those in the full gaze of the public. It recognizes that, in most museums, the majority of the menagerie is housed behind the scenes. This was the result of an increasing separation of displays intended for general audiences from collections reserved for students and specialists, a dichotomy which arose during the latter part of the nineteenth century.[15] William Henry Flower was one of the principal advocates for the partitioning of collections according to assumptions of their potential use:

> The possessions of a great museum of natural history must be divided into two distinct parts—to be separately dealt with in almost all respects —viz. the public or show collection, and the special or study-collection, not exhibited to the general public, but readily accessible to all investigators and specially-qualified persons. The latter collection, [Flower] insisted, should have at least as much space devoted to it as the former. In this way the public galleries would (he showed) be cleared of the excess of specimens which, nonetheless, the museum must carefully preserve for the use of specialists.[16]

Yet, the dichotomy between front of house and behind the scenes, between visible and other spaces, is not as fixed and rigid as might be assumed. Animal representations pass back and forth between these spaces, being prepared behind the scenes and possibly returning there for conservation or repair. Hides at one time forming cabinet collections may be mounted for display, or mounts relaxed and converted into "flat-pack" cabinet skins. The behind-the-scenes spaces have their own visitors, albeit their visits are generally more sporadic and subject to tighter controls and scrutiny. Yet these visits are often more intense and more intimate, involving not only visual engagement between visitor and specimen, but also tactile encounters. It would, therefore, be a mistake to consider only the front-of-house spaces of the museum as containing items that are on display. The behind-the-scenes objects, too, are "displayed," typically through published catalogues and scientific publications, to attract and engage audiences, albeit different audiences from those to whom objects are exhibited in the front of house. Sophie Everest, in her discussion of the Manchester Mandrill in this volume, notes that while these backstage places may be subject to only sporadic visits and long periods of apparent inactivity, these, too, contain their own dynamic suggestive of changing status, values, and meanings. There is no need here to demonstrate further that animals of all kinds perform a multiplicity of afterlife roles, both on and off display. Therefore, I now turn to the technologies which preserve, maintain, and (re)present the material forms which constitute the museum menagerie.

## THE LOOK OF LIFE

Central to the consideration of afterlife in the earlier essays have been the products of taxidermy; a technology or suite of craft processes by which a carcass is transformed into a particular kind of museum object. In this section, I explore taxidermy, and other preservation technologies and techniques, in relation to representational practices of museums.

Typically taxidermy involves preserving the external surface layers of the carcass by "cleaning" away those deeper subdermal parts likely to act as a substrate for fungi or bacteria. The skin is then rendered inaccessible to agents of decay by desiccation or chemical tanning. This produces a remnant which is stable in form (although not in meaning). The general form of the discarded parts of the carcass is then reconstructed by processes of sculpting and modeling, with the skin being incorporated as the outer surface of the model. Taxidermy has been defined in terms of the intention to prepare a specimen so as to assuage and efface death and "to impart to a shapeless skin the exact size, the form, the attitude, the look of life."[17] It is these re-cycled and re-presented structures of biological life, which efface the work put into their production, that for some, exert affective power: "Taxidermy is an encounter between you [and] an animal. It is a strangely and deeply intimate encounter."[18]

However, defining taxidermy as aiming to create "the look of life" situates it as only one of the set of practices undertaken by individuals and firms styling themselves as "taxidermists." A variety of other products emanate (and have emanated) from taxidermists' workshops—prepared skeletons, either articulated or disarticulated, mounted horns, tusks, or heads as trophies, cabinet or study skins, and a miscellany of rugs, chairs, ashtrays, wastebaskets, standard lamps, and other "animal furniture" (fig. 1).[19] Despite the variety of their products, many taxidermists sought to establish identities which distinguished themselves from those of other preparators whose practices did not include elements of model making. As one taxidermist wrote, "Embalming as simply a means of preserving is a separate art, and cannot, strictly speaking, come under the head of taxidermy, while taxidermy proper attempts to reproduce the forms, attitudes and expressions of animals as they appear in life."[20] Nonetheless, practices of fluid preservation and those of modeling, or the products of these practices, coexisted in many museums, and taxidermists were just part of the network of "merchant naturalists" who, directly or through marketing agents, stocked the museum menagerie.[21] The commercial activities of these merchants—supplying animal representations, pinned insects and other desiccated animals, and jars of fluid-preserved animal carcasses—supplemented the material acquired by museums from collectors. Thus the museum menagerie may be constituted from a diversity of entities; some whole animals desiccated or fluid-preserved, others models or reconstructions fashioned to resemble the animal in life. Particularly in the twentieth century, with the increasing distinction between display items and research collections, many items, including most of the fluid-preserved collections, were relocated to see out their afterlives behind the scenes—in most museums, by far the majority of the collections are now behind the scenes. Yet, whether on public exhibition or not, animal remains perform various afterlife roles in forging understandings of the natural world.

In the museum, organic and inorganic creatures rub shoulders. While taxidermy is a form of remnant modeling, taxidermists have also engaged in modeling practices which incorporate none of the original animal. They include molding and casting of fish and live modeling of a range of vertebrates from amphibians to humans, creating impressions both literally and metaphorically.[22] These techniques are sufficiently widespread to prompt one recent ex-

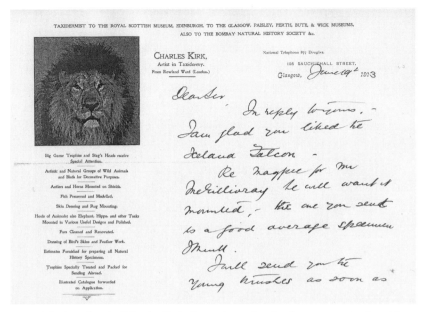

FIG. 1. The letterhead of Charles Kirk, from a letter dated 19 June 1913 to the Scottish naturalist J. A. Harvie Brown. Kirk's stationery indicates the range of work undertaken by his firm. (National Museums Scotland Library Archive, Harvie Brown Collection 28/467; © National Museums Scotland)

pansion of the definition of the craft: "Taxidermy is a general term describing the many methods of reproducing a life-like three-dimensional representation of an animal for permanent display. In some cases, the actual skin (including the fur, feathers or scales) of the specimen is preserved and mounted over an artificial armature. In other cases, the specimen is reproduced completely with man-made materials."[23] Other model-making practices, such as sculpting biological models in wax or glass, or the manufacture of glass eyes for insertion into pieces of taxidermy, are associated with specialist model makers who form other parts of the network of naturalist merchants. Some inhabitants of the museum menagerie are the work of these (nontaxidermist) modelers, in which nonanimal components mimic biological entities. As Henry Augustus Ward, who operated his biological supply company in Rochester, New York, wrote in his marketing of the glass models made in Saxony by Leopold Blaschka (fig. 2):

Everyone familiar with the living sea-anemones, polyps, and jellyfish, knows how utterly impossible it is to preserve either their form or colour after death. Yet without these forms, represented in some way, it is impossible to get or convey any complete or symmetrical knowledge of invertebrate zoology. Alcoholic preparations are invaluable for dissection and special anatomical study, but what is to replace in these groups . . . the stuffed skins and prepared skeletons with which we illustrate the groups of higher animals?[24]

In the nineteenth century and into the early decades of the twentieth century, casts and accurate models were widely considered crucial to the educational role of a museum.[25] The use of casts, models (both remnant and nonremnant), electrotypes, and other copies or simulacra as surrogates for specimens was widespread. As Cyrus Adler observed, "for the purposes of study a cast was as good as an original."[26] Consequently, there was "the willing acceptance of copies, casts, impressions . . . and other surrogates for primary artifacts amongst museum professional and museum visitor alike."[27] Nonremnant models of all kinds, ranging from small-scale representations of dinosaurs to enlargements of microscopic organisms, were acquired by museums.[28] These modeled representations were somehow more real than the illustrations on which many were modeled. Given the stamp of authority and authenticity by their inclusion in the museum, they vouched for the superficial appearance of the biological entities on which they were based. Like the remnant models, these models form part of the circulation of "knowledge in transit" which is productive of views of the natural world. Animals in indexical form, as cast or model, may also be considered to perform roles in afterlife. As some of the other essays in this collection have noted, other indexes of animal photographs, cartoons, postcard, and computer images circulate beyond the museum, extending the afterlife roles beyond the confines of the museum building.

CONCLUSION

This collection of essays demonstrates that while museum collections largely comprise biologically dead animals, this "abundant harvest yielded by death" is as much material culture as other kinds of collections.[29] My objective in this afterword has been to supplement and expand the concept of the museum menagerie, and that of the afterlives it vitalizes, beyond the areas covered in

FIG. 2. Models of marine and terrestrial invertebrates, made by Leopold Blaschka, ca. 1866–68. (© National Museums Scotland)

the preceding essays. The postmortem reanimation of once biologically living objects as material culture is a process of narration. Afterlives are created through their telling—lives told, lives no longer (biologically) lived.[30] The museum menagerie may be narrated in a variety of ways and in a variety of sites, which are not confined to the exhibition spaces of the museum.

The preceding essays have considered individual afterlives, but the museum menagerie may also perform collectively. In the mid-nineteenth century, Edward Forbes considered that the arrangement and juxtaposition of components of museum taxonomic displays provided lessons in civics; and their viewing involved what Tony Bennett dubbed "civic seeing."[31] The arrangement of zoological displays, in Forbes's view, was calculated to create in the workingman a "reverential sense of the extent of knowledge possessed by his fellow-men." This was a function, not of the individual objects, but of their relationship, one to another: "It is not the objects themselves that he [the workingman] sees there and wonders at, that make this impression, so much as the order and evident science which he cannot but recognize in the manner in which they are grouped and arranged." Forbes believed that these visual lessons would foster a "thirst for natural knowledge, one which promised to quench the thirst for beer and vicious excitement." They had the power to make the workingman "a better citizen and happier man" through his appreciation of his proper place in the natural order of the world—a place from which he should not seek to stray.[32] Thus, for Forbes, the ordered arrangement of the museum menagerie (as a whole) had vitality in directing (human) social structures.

It is engagement in the process of narration and meaning-making, not the material form of the animal representation, which constitutes an afterlife. Accession into a museum collection is a rite of passage—the ritual of accessioning an individual animal into a museum grants that individual a museum afterlife. Conceptualized in this way, all the inhabitants of the museum menagerie are vital cultural objects which in diverse ways, in material and indexical form, enact a multiplicity of afterlives within and beyond the museum. It is through the stories that we tell about these animals that they spring into afterlife.

NOTES

1. Ruthven Tremain, *The Animals' Who's Who: 1,146 Celebrated Animals in History, Popular Culture and Lore* (London: Routledge, 1982).

2. "Pete the Pilot" was based on an initial misidentification of the animal as a pilot whale, *Globicephala* sp.; "Willy" made reference to the 1993 feature film *Free Willy*, and "Wally" is a slang term for someone clumsy or foolish, the sort who might get disorientated and lost. The headline "Celebrity Big Blubber," which appeared in the newspaper the *Sun* (21 January 2006), was a reference to the reality series *Celebrity Big Brother* then showing on British television.

3. Marius Kwint, "Introduction: The Physical Past," in *Material Memories,* ed. Kwint, Christopher Breward, and Jeremy Aynsley (Berg: Oxford, 1999), 1–16.

4. Susan Leigh Star, "Craft vs. Commodity, Mess vs. Transcendence: How the Right Tool Became the Wrong One in the Case of Taxidermy and Natural History," in *The Right Tools for the Job: At Work in Twentieth-Century Life Sciences,* ed. Adele E. Clarke and Joan H. Fujimura (Princeton: Princeton University Press, 1992), 257–86. For a discussion of preparation as cleaning, see Samuel J. M. M. Alberti, *Nature and Culture: Objects, Disciplines and the Manchester Museum* (Manchester: Manchester University Press, 2009).

5. Examples from popular literature include: *Little Encyclopedia of Animals, Birds, Fish, Reptiles and Insects* (London: Tyndall, 1968); Barbara Cork, Rosamund K. Cox, and Alwynne C. Wheeler, *Animals, Birds and Fishes* (London: Usbourne, 1990); Alan C. Jenkins, *Wildlife in the City: Animals, Birds, Reptiles, Insects and Plants in an Urban Landscape* (Exeter: Webb and Bower, 1982); and Celia Shute, *Painting Animals and Birds on China* (Wakefield: Westfield House, 2002). Recent examples from the specialist literature include: Yoav Bashan, "Field Dispersal of *Pseudomonas syringae* pv. *tomato, Xanthomonas campestris* pv. *vesicatoria,* and *Alternaria macrospora* by Animals, People, Birds, Insects, Mites, Agricultural Tools, Aircraft, Soil Particles, and Water Sources," *Canadian Journal of Botany* 64 (1986): 276–81; and Barry R. Blakley and Margaret J. Yole, "Species Differences in Normal Brain Cholinesterase Activities of Animals and Birds," *Veterinary and Human Toxicology* 44, no. 3 (2002): 129–32.

6. *Philip Marsden Introduces Sparkie Williams* (Parlaphone 45–R 4475, 1958); *Bird Mimicry: A Remarkable Collection of Imitations by Birds* (British Library Publishing Division, 2006); Carsten Nicolai [Alva Noto] and Michael Nyman, *Sparkie: Cage and Beyond,* performance, March 2009, Berlin; Michael Nyman, *Pretty Talk for George Brecht,* performance, ca. 1977. The Hancock Museum is now part of the Great North Museum.

7. James Meikle, "Anglers Mourn Benson the 29kg Fish: Greedy Carp the Size of a Dog Thought to Have Died from Overdose of Uncooked Nuts," *Guardian,* 4 August 2009; Henry Nicholls, *Lonesome George: The Life and Loves of a Conservation Icon* (London: Pan, 2007); Christophe Nkurunziza, "Burundi's Not So Gentle Giant," *BBC News,* 29 November 2002, http://news.bbc.co.uk/1/hi/world/africa/2520815 .stm; Col Stringer, *The Saga of Sweetheart: The Frightening but True Story of the Giant Rogue Crocodile who Attacked over 15 Boats on a N.T. River during the 1970s* (Casuarina, N.T. [Australia]: Adventure, 1986); Chris Ryan, "Saltwater Crocodiles as Tourist Attractions," *Journal of Sustainable Tourism* 6 (1998): 314–27.

8. Geoffrey N. Swinney, "Granny (*c.* 1821–1887), 'A Zoological Celebrity,'" *Archives of Natural History* 34 (2007): 219–28; Anon., "World Cup Oracle Octopus Paul Dies," CNN, 26 October 2010, http://edition.cnn.com/2010/SPORT/10/26/ germany.paul.octopus.death/index.html; David Crossland, "Death of an Oracle: Rest in Peace, Paul the Octopus," *Spiegel Online International,* 26 October 2010, www.spiegel.de/international/zeitgeist/0,1518,725399,00.html; Roger Boyes, "RIP Paul the Oracle Octopus: His Demise Was One Result That He Did Not See Coming," *Times* (Scotland edition), 27 October 2010; Tania Branigan, Kate Connelly, and Sam Jones, "Kicker Conspiracy? Director Casts Doubt on Death of Psychic Octopus," *Guardian,* 27 October 2010.

9. "Dolly the Sheep," in *Connect: Creativity, Innovation and Discovery* (Edinburgh: NMSE, 2006), 10–11; Ian Wilmut, Angelika E. Schnieke, James McWhir, Alexander J. Kind, and Keith H. S. Campbell, "Viable Offspring Derived from Fetal and Adult Mammalian Cells," *Nature* 385, no. 6619 (1997): 810–13.

10. For a discussion of the nature of museum documentation, see Jenny Walklate, "The Quotidian and the Bizarre: Rewriting the Day Book" (master's thesis, University of Leicester, 2009); Geoffrey N. Swinney, "What Do We Know about What We Know?—The Museum 'Register' as Museum Object," in *The Thing about Museums: Objects and Experience, Representation and Contestation,* ed. Sandra Dudley et al. (London: Routledge, 2011).

11. Rosamond W. Purcell and Stephen Jay Gould, *Illuminations: A Bestiary* (New York: Norton, 1986), 116.

12. Carol K. Yoon, *Naming Nature: The Clash between Instinct and Science* (New York: Norton, 2009); International Commission on Zoological Nomenclature, *International Code of Zoological Nomenclature,* 4th ed. (2000), www.iczn.org/iczn/index.jsp.

13. John T. Huber, "The Importance of Voucher Specimens, with Practical Guidelines for Preserving Specimens of the Major Invertebrate Phyla for Identification," *Journal of Natural History* 32 (1998): 367–85.

14. Igor Kopytoff, "The Cultural Biography of Things: Commoditization as a Process," in *The Social Life of Things: Commodities in Cultural Perspective,* ed. Arjun Appadurai (Cambridge: Cambridge University Press, 1986), 64–91.

15. Thomas Henry Huxley, "Suggestions for a Proposed Natural History Museum in Manchester," *Report of the Proceedings of the Museums Association* 7 (1896 [from a 1868 manuscript]): 126–31; William H. Flower, "Museum Organisation," in *Essays on Museums and Other Subjects Connected with Natural History* (London: Macmillan, 1898), 1–29; Ramsay H. Traquair, "Presidential Address: Some Points in Connection with Natural History Museums," *Proceedings of the Royal Physical Society of Edinburgh* 11 (1892): 173–79;: Francis A. Bather, "Museums Association, The Aberdeen Conference, 1903: Address by the President, Francis Arthur Bather," *Museums Journal* 3 (1903): 71–94.

16. Edwin Ray Lankester, "William Henry Flower," *Nature,* 13 July 1899, 254.

17. Oliver Davie, *Methods in the Art of Taxidermy* (Philadelphia: McKay, 1894), ii; William Temple Hornaday, "Common Faults in the Mounting of Quadrupeds," *Annual Report of Society of American Taxidermists* (1883): 67.

18. Rachel Poliquin, *Ravishing Beasts: The Strangely Alluring World of Taxidermy* (Vancouver: Museum of Vancouver, 2009)

19. For examples of the range of products of one firm, see Pat A. Morris, *Rowland Ward: Taxidermist to the World* (Ascot: MPM, 2003).

20. Davie, *Methods in the Art of Taxidermy,* ii.

21. Sally Gregory Kohlstedt, "Henry A. Ward: Merchant Naturalist and American Museum Development," *Journal of the Society for the Bibliography of Natural History* 9 (1980): 647–61; Kohlstedt, "Museums on Campus: A Tradition of Inquiry and Teaching," in *The American Development of Biology,* ed. Ronald Rainger, Keith R. Benson, and Jean Maienschein (Philadelphia: University of Philadelphia Press, 1988), 15–47

22. Edwin Leonard Gill and H. Fletcher, "Plaster Casts of Fishes," *Museums Journal* 14 (1915): 253–59, 289–98; Jim Hall, *The Breakthrough Fish Taxidermy Manual,* 2nd ed. (Munroe, Ga.: Breakthrough, 1988); Edward C. Migdalski, *Fish Mounts and Other Fish Trophies: The Complete Book of Fish Taxidermy* (New York: Wiley, 1981). For a recent overview of taxidermy practice, see Melissa Milgrom, *Still Life: Adventures in Taxidermy* (Boston and New York: Houghton Mifflin Harcourt, 2010).

23. "Taxidermy 101," www.capricorntaxidermy.com/docs/taxidermy%20101.pdf.

24. Henry Augustus Ward, *Catalogue of Glass Models of Invertebrate Animals* (Rochester, N.Y.: Ward's Natural Science Establishment, 1878). For a discussion of Ward's business, see Roswell Ward, *Henry A. Ward: Museum Builder to America,* ed. Blake McKelvey (Rochester, N.Y.: Rochester Historical Society, 1948).

25. Henry Browne, "The Influence of Museums on the Reform of Classical Studies," *Museums Journal* 12 (1913): 197–20; Ben H. Mullen, "Greek Sculpture—Some Notes on Collecting and Exhibiting Plaster Reproductions," *Museums Journal* 12 (1913): 297–304; Geoffrey N. Swinney, "Enchanted Invertebrates: Blaschka Models and Other Simulacra in National Museums Scotland," *Historical Biology* 20 (2008): 39–50.

26. Cyrus Adler, *I Have Considered the Days* (Philadelphia: Jewish Publication Society, 1941), 67.

27. Barbara Kirshenblatt-Gimblett, *Destination Culture: Tourism, Museums, and Heritage* (Berkeley and Los Angeles: University of California Press, 1998), 31.

28. For further discussion on the use of models and casts, see Soraya de Chadarevian and Nick Hopwood, eds., *Models: The Third Dimension of Science* (Stanford, Calif.: Stanford University Press, 2004).

29. The quotation is from Frederick Knight Hunt, "The Hunterian Museum," *Household Words,* 14 December 1850, 279.

30. Trevor J. Barnes, "Lives Lived and Lives Told: Biographies of Geography's Quantitative Revolution," *Environment and Planning D: Society and Space* 19 (2001): 409–29.

31. Tony Bennett, "Civic Seeing: Museums and the Organization of Vision," in *A Companion to Museum Studies,* ed. Sharon Macdonald (Oxford: Blackwell, 2006), 263–81; Geoffrey N. Swinney, "Edward Forbes (1815–1854) and the Exhibition of Natural Order in Edinburgh," *Archives of Natural History* 37 (2010): 309–17.

32. Edward Forbes, *Museum of Practical Geology: On the Educational Uses of Museums* (London: Longman, 1853), 9, 15.

Alberti, Samuel J. M. M. "Constructing Nature behind Glass." *Museum and Society* 6 (2008): 73–97.
———. "Objects and the Museum." *Isis* 96 (2005): 559–71.
Altick, Richard D. *The Shows of London: A Panoramic History of Exhibitions, 1600–1862.* Cambridge: Belknap Press of Harvard University Press, 1978.
The Animal Studies Group. *Killing Animals.* Urbana: University of Illinois Press, 2006.
Asma, Stephen T. *Stuffed Animals & Pickled Heads: The Culture and Evolution of Natural History Museums.* Oxford and New York: Oxford University Press, 2001.
Baker, Steve. *The Postmodern Animal.* London: Reaktion, 2000.
Baratay, Eric, and Elisabeth Hardouin-Fugier. *Zoo: A History of Zoological Gardens in the West.* Translated by Oliver Welsh. London: Reaktion, 2002.
Barringer, Tim, and Tom Flynn, eds. *Colonialism and the Object: Empire, Material Culture, and the Museum.* London: Routledge, 1998.
Bennett, Tony. *The Birth of the Museum: History, Theory, Politics.* London: Routledge, 1995.
———. *Pasts beyond Memory: Evolution, Museums, Colonialism.* London: Routledge, 2004.
Berger, John. "Why Look at Animals?" In *About Looking,* by Berger, 1–28. New York: Pantheon, 1980.
Black, Graham. *The Engaging Museum: Developing Museums for Visitor Involvement.* Abingdon: Routledge, 2001.
Blackwell, Mark, ed. *The Secret Life of Things: Animals, Objects, and It-Narratives in Eighteenth-Century England.* Lewisburg: Bucknell University Press, 2007.
Brantz, Dorothee, ed. *Beastly Natures: Animals, Humans, and the Study of History.* Charlottesville: University of Virginia Press, 2010.
Chadarevian, Soraya de, and Nick Hopwood, eds. *Models: The Third Dimension in Science.* Stanford: Stanford University Press, 2004.
Croke, Vicki. *The Modern Ark: The Story of Zoos: Past, Present, and Future.* New York: Scribner, 1998.
Dahlbom, Taika. "A Mammoth History: The Extraordinary Journey of Two Thighbones." *Endeavour* 31 (2007): 110–14.
———. "Matter of Fact: Biographies of Zoological Specimens." *Museum History Journal* 2 (2009): 51–72.
Daston, Lorraine, ed. *Biographies of Scientific Objects.* Chicago: University of Chicago Press, 2000.
———, ed. *Things That Talk: Object Lessons from Art and Science.* New York: Zone, 2004.

Daston, Lorraine, and Peter Galison. *Objectivity.* Boston: Zone, 2007.

Daston, Lorraine, and Gregg Mitman, eds. *Thinking with Animals: New Perspectives on Anthropomorphism.* New York: Columbia University Press, 2005.

DeSilvey, Caitlin. "Observed Decay: Telling Stories with Mutable Things." *Journal of Material Culture* 11 (2006): 318–38.

Desmond, Jane. "Postmortem Exhibitions: Taxidermied Animals and Plastinated Corpses in the Theaters of the Dead." *Configurations* 16 (2008): 347–77.

Elsner, John, and Roger Cardinal, eds. *The Cultures of Collecting.* London: Reaktion, 1994.

Falk, John Howard, and Lynn Diane Dierking. *The Museum Experience.* Washington, D.C.: Howells House, 1992.

Farber, Paul Lawrence. "The Development of Taxidermy and the History of Ornithology." *Isis* 68 (1977): 550–66.

Flower, William Henry. *Essays on Museums and Other Subjects Connected with Natural History.* London: Macmillan, 1898.

Fortey, Richard. *Dry Store Room No. 1: The Secret Life of the Natural History Museum.* London: Harper, 2008.

Foster, Kate, and Hayden Lorimer. "Some Reflections on Art-Geography as Collaboration." *Cultural Geographies* 14 (2007): 425–32.

Fudge, Erica. *Animal.* London: Reaktion, 2002.

Fyfe, Aileen, and Bernard Lightman, eds. *Science in the Marketplace: Nineteenth-Century Sites and Experiences.* Chicago: University of Chicago Press, 2007.

Ghiselin, Michael T., and Alan E. Leviton, eds. *Cultures and Institutions of Natural History: Essays in the History and Philosophy of Science.* San Francisco: California Academy of Sciences, 2000.

Gosden, Chris, and Yvonne Marshall. "The Cultural Biography of Objects." *World Archaeology* 31 (1999): 169–78.

Greenblatt, Stephen. "Resonance and Wonder." In *Exhibiting Cultures: The Poetics and Politics of Museum Display,* edited by Ivan Karp and Steven D. Lavine, 42–56. Washington, D.C.: Smithsonian Institution Press, 1991.

Griesemer, James. "Modelling in the Museum: On the Role of Remnant Models in the Work of Joseph Grinnell." *Biology and Philosophy* 5 (1990): 3–36.

Gutwill-Wise, Josh, and Sue Allen. "Finding Significance: Testing Methods for Encouraging Meaning-Making in a Science Museum." *Current Trends in Audience Research and Evaluation* 15 (2002): 5–11.

Haraway, Donna Jeanne. *Primate Visions: Gender, Race and Nature in the World of Modern Science.* London: Routledge, 1989.

Henare, Amiria, Martin Holbraad, and Sari Wastell, eds. *Thinking through Things: Theorising Artefacts Ethnographically.* London: Routledge, 2006.

Henning, Michelle. "Anthropomorphic Taxidermy and the Death of Nature: The Curious Art of Hermann Ploucquet, Walter Potter, and Charles Waterton." *Victorian Literature and Culture* 35 (2007): 663–78.

Hoage, R. J., and William A. Deiss, eds. *New Worlds, New Animals: From Menagerie to Zoological Park in the Nineteenth Century.* Baltimore: Johns Hopkins University Press, 1996.

Hooper-Greenhill, Eilean. *Museums and the Interpretation of Visual Culture.* London: Routledge, 2000.

Hoskins, Janet. *Biographical Objects: How Things Tell the Stories of People's Lives.* New York: Routledge, 1998.

Ingold, Tim, ed. *What Is an Animal?* London: Routledge, 1988.

Kalof, Linda, and Amy Fitzgerald. "Reading the Hunting Trophy: Exploring the Display of Dead Animals in Hunting Magazines." *Visual Studies* 18 (2003): 112–22.

Kalof, Linda, and Brigitte Resl, eds. *A Cultural History of Animals.* 6 vols. Oxford: Berg, 2007.

Kerasote, Ted. *Bloodties: Nature, Culture, and the Hunt.* New York: Kodansha International, 1993.

Kirshenblatt-Gimblett, Barbara. *Destination Culture: Tourism, Museums and Heritage.* Berkeley and Los Angeles: University of California Press, 1998.

Koenigsberger, Kurt. *The Novel and the Menagerie: Totality, Englishness, and Empire.* Columbus: Ohio State University Press, 2007.

Kohler, Robert E. *All Creatures: Naturalists, Collectors, and Biodiversity, 1850–1950.* Princeton: Princeton University Press, 2006.

———. *Landscapes and Labscapes: Exploring the Lab-Field Border in Biology.* Chicago: University of Chicago Press, 2002.

Kohlstedt, Sally Gregory. "Masculinity and Animal Display in Nineteenth-Century America." In *Figuring It Out: Science, Gender, and Visual Culture,* edited by Ann B. Shteir and Bernard Lightman, 110–39. Hanover, N.H.: University Press of New England, 2006.

Kopytoff, Igor. "The Cultural Biography of Things: Commoditization as Process." In *The Social Life of Things: Commodities in Cultural Perspective,* edited by Arjun Appadurai, 64–91. Cambridge: Cambridge University Press, 1986.

Kowalsky, Nathan, ed. *Hunting and Philosophy.* Malden, Mass.: Blackwell, 2010.

Kwint, Marius, Christopher Breward, and Jeremy Aynsley, eds. *Material Memories.* Oxford: Berg, 1999.

Latham, Kiersten F. "The Poetry of the Museum: A Holistic Model of Numinous Museum Experiences." *Museum Management and Curatorship* 22 (2007): 247–63.

Leviton, Alan E., and Michele L. Aldrich, eds. *Museums and Other Institutions of Natural History, Past, Present, and Future.* San Francisco: California Academy of Sciences, 2004.

Lippit, Akira Mizuta. *Electric Animal: Toward a Rhetoric of Wildlife.* Minneapolis: University of Minnesota Press, 2000.

Lorimer, Hayden. "Guns, Game and the Grandee: The Cultural Politics of Deer Stalking in the Highlands." *Ecumene* 7 (2000): 403–31.

———. "Herding Memories of Humans and Animals." *Environment and Planning D: Society and Space* 24 (2006): 497–518.

Lorimer, Jamie. "Nonhuman Charisma." *Environment and Planning D: Society and Space* 24 (2007): 911–32.

MacGregor, Arthur. *Curiosity and Enlightenment: Collectors and Collections from the Sixteenth to the Nineteenth Century*. New Haven: Yale University Press, 2007.

Machin, Rebecca. "Gender Representation in the Natural History Galleries at the Manchester Museum." *Museum and Society* 6 (2008): 54–67.

MacKenzie, John M. *Museums and Empire: Natural History, Human Cultures and Colonial Identities*. Manchester: Manchester University Press, 2009.

Malamud, Randy. *Reading Zoos: Representations of Animals and Captivity*. New York: New York University Press, 1998.

Maroević, Ivo. "The Museum Message: Between the Document and Information." In *Museum, Media, Message,* edited by Eilean Hooper-Greenhill, 24–36. London: Routledge, 1995.

Midgley, Mary. *Beast and Man: The Roots of Human Nature*. London: Methuen, 1979.

Milgrom, Melissa. *Still Life: Adventures in Taxidermy*. New York: Houghton Mifflin Harcourt, 2010.

Mitchell, W. J. Thomas. *The Last Dinosaur Book: The Life and Times of a Cultural Icon*. Chicago: University of Chicago Press, 1998.

Mitman, Gregg. *Reel Nature: America's Romance with Wildlife on Film*. Cambridge: Harvard University Press, 1999.

Morris, Pat A. "An Historical Review of Bird Taxidermy in Britain." *Archives of Natural History* 20 (1993): 241–55.

———. *Rowland Ward: Taxidermist to the World*. Ascot: MPM, 2003.

Morse, Deborah Denenholz, and Martin Danahay, eds. *Victorian Animal Dreams: Representations of Animals in Victorian Literature and Culture*. Aldershot: Ashgate, 2007.

Mullan, Bob, and Garry Marvin. *Zoo Culture: The Book about Watching People Watch Animals*. 2nd ed. Urbana: University of Illinois Press, 1999.

Nicholls, Henry. *Lonesome George: The Life and Loves of a Conservation Icon*. London: Pan, 2007.

———. *The Way of the Panda: The Curious History of China's Political Animal*. London: Profile, 2010.

Paddon, Hannah. "An Investigation of the Key Factors and Processes That Underlie the Contemporary Display of Biological Collections in British Museums." Ph.D. diss., Bournemouth University, 2010.

Paris, Scott G., ed. *Perspectives on Object-Centered Learning in Museums*. Mahwah, N.J.: Lawrence Erlbaum, 2002.

Patchett, Merle. "Putting Animals on Display: Geographies of Taxidermy Practice." Ph.D. diss., University of Glasgow, 2009.

———. "Tracking Tigers: Recovering the Embodied Practices of Taxidermy." *Historical Geography* 36 (2008): 98–122.

Patchett, Merle, and Kate Foster. "Repair Work: Surfacing the Geographies of Dead Animals." *Museum and Society* 6 (2008): 98–122.

Philo, Chris, and Chris Wilbert, eds. *Animal Spaces, Beastly Places: New Geographies of Human-Animal Relations.* London: Routledge, 2000.

Plumb, Christopher. "Exotic Animals in Eighteenth-Century Britain." Ph.D. diss., University of Manchester, 2010.

———. "'In Fact, One Cannot See It without Laughing': The Spectacle of the Kangaroo in London, 1770–1830." *Museum History Journal* 3 (2010): 7–32.

Poliquin, Rachel. "The Matter and Meaning of Museum Taxidermy." *Museum and Society* 6 (2008): 123–34.

———. *Ravishing Beasts: The Strangely Alluring World of Taxidermy.* Exhibition catalogue. Vancouver: Museum of Vancouver, 2009.

Prince, Sue Ann, ed. *Stuffing Birds, Pressing Plants, Shaping Knowledge: Natural History in North America, 1730–1860.* Philadelphia: American Philosophical Society, 2003.

Purcell, Rosamond Wolff, and Stephen Jay Gould. *Illuminations: A Bestiary.* New York: Norton, 1986.

Quinn, Stephen C. *Windows on Nature: The Great Habitat Dioramas of the American Museum of Natural History.* New York: Abrams, 2006.

Rader, Karen A., and Victoria E. M. Cain. "From Natural History to Science: Display and the Transformation of American Museums of Science and Nature." *Museum and Society* 6 (2008): 152–71.

Riggins, Stephen Harold, ed. *The Socialness of Things: Essays on the Socio-Semiotics of Objects.* Berlin and New York: Mouton de Gruyter, 1994.

Ritvo, Harriet. *The Animal Estate: The English and Other Creatures in the Victorian Age.* London: Penguin, 1987.

———. *Noble Cows and Hybrid Zebras: Essays on Animals and History.* Charlottesville: University of Virginia Press, 2010.

———. *The Platypus and the Mermaid and Other Figments of the Classifying Imagination.* Cambridge and London: Harvard University Press, 1997.

Robbins, Louise E. *Elephant Slaves and Pampered Parrots: Exotic Animals in Eighteenth-Century Paris.* Baltimore: Johns Hopkins University Press, 2002.

Rossi, Michael. "Fabricating Authenticity: Modeling a Whale at the American Museum of Natural History, 1906–1974." *Isis* 101 (2010): 338–61.

Rothfels, Nigel, ed. *Representing Animals.* Bloomington and Indianapolis: Indiana University Press, 2002.

———. *Savages and Beasts: The Birth of the Modern Zoo.* Baltimore: Johns Hopkins University Press, 2002.

Scott, Monique. *Rethinking Evolution in the Museum: Envisioning African Origins.* London: Routledge, 2007.

Secord, James A. "Knowledge in Transit." *Isis* 95 (2004): 654–72.

Snæbjörnsdóttir, Bryndís, and Mark Wilson, *Nanoq: Flatout and Bluesome: A Cultural Life of Dead Animals.* London: Black Dog, 2006.

Spencer, Tom, and Sarah Whatmore. "Bio-Geographies: Putting Life Back into the Discipline." *Transactions of the Institute of British Geographers* 26 (2001): 139–41.

Star, Susan Leigh. "Craft Vs. Commodity, Mess Vs. Transcendence: How the Right Tool Became the Wrong One in the Case of Taxidermy and Natural History." In *The Right Tools for the Job: At Work in Twentieth-Century Life Sciences,* edited by Adele E. Clarke and Joan H. Fujimura, 257–86. Princeton: Princeton University Press, 1992.

Star, Susan Leigh, and James R. Griesemer. "Institutional Ecology, 'Translations' and Boundary Objects: Amateurs and Professionals in Berkeley's Museum of Vertebrate Zoology, 1907–39." *Social Studies of Science* 19 (1989): 387–420.

Stearn, William Thomas. *The Natural History Museum at South Kensington: A History of the British Museum (Natural History) 1753–1980.* London: Heinemann, 1981.

Stewart, Susan. *On Longing: Narratives of the Miniature, the Gigantic, the Souvenir, the Collection.* Durham, N.C.: Duke University Press, 1993.

Suarez, Andrew V., and Neil D. Tsutsui. "The Value of Museum Collections for Research and Society." *BioScience* 54 (2004): 66–74.

Swinney, Geoffrey N. "'Granny' (*c.* 1821–1887), 'A Zoological Celebrity.'" *Archives of Natural History* 34 (2007): 219–28.

Thorsen, Liv Emma. "A Fatal Visit to Venice: The Transformation of an Indian Elephant." In *Investigating Human/Animal Relations in Science, Culture and Work,* edited by Tora Holmberg, 85–96. Uppsala: Centrum för genusvetenkap, 2009.

Tremain, Ruthven. *The Animals' Who's Who: 1,146 Celebrated Animals in History, Popular Culture and Lore.* London: Routledge, 1982.

Wakeham, Pauline. *Taxidermic Signs: Reconstructing Aboriginality.* Minneapolis: University of Minnesota Press, 2008.

Winker, Kevin. "Natural History Museums in a Postdiversity Era." *BioScience* 54 (2004): 455–59.

Wonders, Karen. "Habitat Dioramas and the Issue of Nativeness." *Landscape Research* 28, no. 1 (2003): 89–100.

———. *Habitat Dioramas: Illusions of Wilderness in Museums of Natural History.* Uppsala: Almqvist and Wiksell, 1993.

Yanni, Carla. *Nature's Museums: Victorian Science and the Architecture of Display.* London: Athlone, 1999.

SAMUEL J. M. M. ALBERTI is the Director of Museums and Archives at the Royal College of Surgeons of England and a Visiting Senior Research Fellow at King's College London. He is the author of *Nature and Culture: Objects, Disciplines and the Manchester Museum* (2009) and *Morbid Curiosities: Medical Museums in Nineteenth-Century Britain* (2011).

SOPHIE EVEREST has worked as a television producer and journalist for the BBC, and is now researching the history of taxidermy.

KATE FOSTER is an environmental artist who works on and around human and animal lives in an unsteady world. Her work includes *Out of Time* (2007). She has collaborated with Hayden Lorimer and Merle Patchett around the theme of biogeographies.

MICHELLE HENNING is a writer, artist, and cultural historian. She is a Senior Lecturer in Media and Cultural Studies in the Faculty of Creative Arts at the University of the West of England, Bristol, and the author of *Museums, Media and Cultural Theory* (2006).

HAYDEN LORIMER is a Senior Lecturer at the University of Glasgow Department of Geographical and Earth Sciences. He edited *Practising the Archive: Reflections on Method and Practice in Historical Geography* (2007). He has collaborated with Kate Foster and Merle Patchett around the theme of biogeographies.

GARRY MARVIN is the Professor of Human-Animal Studies at Roehampton University, London. His books include *Bullfight* (new ed., 1994) and *Zoo Culture* (new ed., 1999).

HENRY NICHOLLS is a freelance science journalist. He is the author of *Lonesome George: The Life and Loves of the World's Most Famous Tortoise* (2007) and *The Way of the Panda: The Curious History of China's Political Animal* (2010).

HANNAH PADDON recently completed her Ph.D. on the contemporary display of biological collections in British museums at the School of Conservation Sciences at Bournemouth University.

MERLE PATCHETT completed her Ph.D. concerning the geographies of taxidermy practice at the University of Glasgow Department of Geographical and Earth Sciences. She has collaborated with Kate Foster and Hayden Lorimer around the theme of biogeographies.

CHRISTOPHER PLUMB taught at the University of Manchester Centre for Museology. Previously he was a Pre-doctoral Research Fellow at the Max Planck Institute for the History of Science in Berlin. He recently completed his Ph.D. on eighteenth-century animals at the University of Manchester.

RACHEL POLIQUIN is a writer and curator, currently working on all things taxidermy. Her forthcoming book is a cultural exploration of taxidermy from cabinets of curiosity to contemporary animal art. She also maintains the taxidermy blog ravishingbeasts.com.

JEANNE ROBINSON is the Curator for Entomology for Culture and Sport Glasgow (Glasgow Museums), and with Mike Rutherford and Richard Sutcliffe was part of the team responsible for launching the renovated Kelvingrove Museum.

MIKE RUTHERFORD, formerly the Curator for Invertebrate Zoology for Culture and Sport Glasgow (Glasgow Museums), is now the Zoology Curator at The University of the West Indies, and with Jeanne Robinson and Richard Sutcliffe was part of the team responsible for launching the renovated Kelvingrove Museum.

RICHARD C. SABIN is the Senior Curator in the Mammal Group at the Natural History Museum and a specialist advisor to the Cetacean Strandings Project. He has published widely on archaeozoology, whales, and collections management.

RICHARD SUTCLIFFE is the Research Manager for Natural History, and with Jeanne Robinson and Mike Rutherford was part of the team responsible for launching the renovated Kelvingrove Museum.

GEOFFREY N. SWINNEY was for more than thirty years the Principal Curator of Lower Vertebrates in the institution currently known as National Museums Scotland, where he remains an Honorary Research Associate. He is currently undertaking research in the historical geography of museums at the University of Edinburgh.

# INDEX

/